AMERICAN AMMUNITION AND BALLISTICS

5p

AMERICAN AMMUNITION AND BALLISTICS

Edward Matunas

WINCHESTER PRESS

Library of Congress Cataloging in Publication Data

Matunas, Edwards.
    American ammunition and ballistics.

    Includes index.
    1. Ammunition.   2. Cartridges.   3. Ballistics.
I. Title.
TS538.M33   683'.406   79-10921
ISBN 0-87691-290-0

9   8   7   6   5   4   3   2   1

Published by Winchester Press
1421 South Sheridan Drive
Tulsa, Oklahoma 74101

Printed in the United States of America

The author dedicates this work to all those
people in the firearms industry who, over
the past 26 years, have assisted or
helped provide the background which has
enabled the author to put together
American Ammunition and Ballistics.

ACKNOWLEDGMENTS

Particular acknowledgment is given to the Sporting Arms and Ammunition Manufacturers' Institute, Inc. without whose help the ballistics tables and much of the other information contained in this work would not have been possible.

Also, the author acknowledges Winchester-Western and Federal Cartridge Corporation for their valuable assistance in supplying a large number of the photos used in this work.

Particular acknowledgment is paid to Mr. Charles Gent who kindled this author's interest in firearms and ammunition when the author was 12 years old. I doubt if Charlie ever dreamt of where a 12-gauge, full-choke, bolt-action shotgun would lead that skinny young lad.

Contents

CHAPTER ONE
Rimfire Ammunition   *1*

CHAPTER TWO
Rimfire Exterior Ballistics Tables
for Rifles and Handguns   *9*

CHAPTER THREE
Recoil Calculation, an Important Step
in Cartridge or Shell Selection   *17*

CHAPTER FOUR
Choosing a Centerfire Rifle Cartridge   *25*

CHAPTER FIVE
Centerfire Exterior Ballistics Tables
for Rifles   *45*

CHAPTER SIX
Choosing a Centerfire Handgun Cartridge   *79*

CHAPTER SEVEN
Centerfire Exterior Ballistics Tables
for Handguns   *89*

CHAPTER EIGHT
Choosing the Correct Gauge and Load   *96*

CHAPTER NINE
Exterior Ballistics Tables for Shotshells   *116*

CHAPTER TEN
The Do's and Don'ts of Ammunition
Interchangeability   *129*

CHAPTER ELEVEN
Practical Cartridge and Shell Selections   *138*

CHAPTER TWELVE
Reloading Ammunition   *145*

CHAPTER THIRTEEN
New Ammunition Developments   *153*

CHAPTER FOURTEEN
Properties of Sporting Ammunition and
Recommendations for its Storage and
Handling, as Supplied by SAAMI   *173*

CHAPTER FIFTEEN
Recommendations for Storage and Control
of Small-Arms Ammunition for Security
and Law-Enforcement Use, as Supplied by
SAAMI   *179*

CHAPTER SIXTEEN
Historical Exterior Ballistics Tables   *188*

CHAPTER SEVENTEEN
Ammunition and Ballistics Glossary   *202*

CHAPTER EIGHTEEN
Shooting Safety   *214*

Foreword

The author has made every attempt to keep personal opinions from creeping into these pages. From the beginning, every effort was made to make this a reference work that would list only the facts about ammunition as they generally are supported throughout the firearms industry.

It is hoped that this reference work will enhance the readers' overall knowledge of ammunition. Surely the author's knowledge was increased by the preparation of the material. Over the years, the reader will find many reasons to refer back to this volume. It is hoped that he or she will enjoy using the book as much as the author enjoyed preparing it.

The author makes no claim that the warnings or cautions contained in the book are all-inclusive. In fact, he is sure that additional warnings and cautions will continually be required.

The author asks the readers' indulgence for Chapter Eleven wherein it will appear that he has allowed personal preferences to guide his comments. It will be up to the reader to decide if this chapter is fact or opinion.

AMERICAN AMMUNITION AND BALLISTICS

CHAPTER ONE

Rimfire Ammunition

Literally billions of rounds of .22-caliber rimfire ammunition are consumed annually. Most of it is expended plinking and in the hunting of squirrels and rabbits. Small varmints, such as woodchucks, are also eagerly sought by rimfire shooters.

Today almost all of the rimfire ammunition used is of .22 caliber with but a few limited exceptions. However, such was not always the case. One of the most sought-after Union weapons of the Civil War was the Spencer carbine. Its very large magazine capacity and the fixed cartridge it used made it a very desirable weapon. The cartridge used in the Spencer was a rimfire cartridge of approximately .52 caliber. The volcanic pistols made from 1858 to 1860 were chambered for a .30-caliber rimfire cartridge, and the volcanic rifle of the same period was chambered for a .38-caliber rimfire cartridge. The Henry rifles of 1860 to 1861 were chambered for a .44-caliber rimfire cartridge, and the first Winchester, the Model 1866, was chambered for a .44-caliber rimfire cartridge.

The Swiss military weapon of years gone by was chambered for a very large bottleneck rimfire case of .41 caliber. Remington Arms produced a very popular over-and-under derringer pistol of .41 caliber, and quite a few early U.S. sporting rifles were chambered for rimfire cases of .25 and .32 caliber.

Such listings could go on and on; indeed, whole volumes have been written on rimfire cartridges. However, all these cartridges are relics of history and are of little importance to the modern shooter. Even today we have a number of these relics still holding on sufficiently well so that we are obligated to mention them in passing. These include:

.22 Winchester Rimfire.
.22 Winchester Auto
.25 Stevens (Short and Long)
.32 (Short and Long)

All of the foregoing can be used successfully on small game and small varmint to ranges of 50 yards or so. However, firearms chambered for these cartridges are encountered infrequently, and ammunition is even harder to find.

The rimfire cartridges being used today are primarily as follows:

> 5mm Remington Magnum (.20 caliber)
> .22 Short
> .22 Long
> .22 Long Rifle
> .22 Winchester Magnum

There are a number of variations which include standard velocity, high velocity, solid bullets, and hollow-point bullets. One or more of these variations can be found in a single cartridge: for example, a .22 Long Rifle, high velocity, hollow point.

The 5mm Remington Magnum never really caught on, and it might not be unfair to say that, despite its relatively recent introduction, it is already a dying cartridge. The remaining rimfires will undoubtedly be with us for as long as we retain the right to bear arms.

The longevity of the various .22-caliber rimfires is assured by their relatively low cost, very low noise level, good accuracy, and sufficient energy to take small varmint and small game at ranges to 50 yards. Under certain circumstances the ranges can be stretched to 75 or 100 yards. But the use of any rimfire cartridges at ranges beyond 100 yards falls into the stunt category and will result in the loss of a lot of crippled game.

Of course, the ability to practice almost anywhere with a .22 is reason enough to ensure its longevity for another century. In these areas, we have several special oddities of .22 rimfire ammunition—the .22 CB Short and the .22 CB Long. These are nothing more than a .22 Short or Long loaded to a reduced velocity of about 675 fps (feet per second). These cartridges allow (with adequate backstop) the use of the .22 indoors or in the backyards of suburban neighborhoods (where not prohibited by law). The low-velocity projectiles ensure that the noise level will be about on par with a high-velocity air gun and that relatively modest backstops will be adequate. The author's sons have used .22 CB Longs quite successfully on a suburban trap line for a number of years. Little or no pelt damage is caused by the low-velocity projectiles.

The long version of these CB rounds is preferred over the

short. By operating the action manually, the long case will function through almost all repeating firearms and even through semiautomatic firearms. Additionally, the longer case will prevent a chamber-erosion problem that could occur with extensive use of the shorter case. See Chapter Ten for further information on this topic.

Complete ballistic tables for all currently manufactured rimfire cartridges follow in Chapter Two. For this chapter we will confine our discussion to the practical selection and application of the various rimfire cartridges being manufactured today.

22 SHORT

The 22 Short is available in a number of variations which are as follows:

.22 CB Short
.22 Short Disintegrating (light bullet)
.22 Short High Velocity Disintegrating (light bullet)
.22 Short Disintegrating (heavy bullet)
.22 Short Standard Velocity
.22 Short High Velocity
.22 Short High Velocity Hollow Point

The .22 CB Short, the .22 Short Disintegrating with 15-grain bullet, the .22 Short High-Velocity Disintegrating with 15-grain bullet, and the .22 Short Disintegrating with 29-grain bullet are suitable only for informal target shooting, plinking, and shooting-gallery applications. Based on industry trends, the disintegrating bullet styles are already doomed. At least one manufacturer plans to discontinue any further production of these rounds.

The .22 Short Standard Velocity with the standard lead bullet is the ideal plinking round. Its low cost, good accuracy, and low noise level add up to the ideal plinker's cartridge. However, there is one drawback. Extensive use of the Short case in a firearm chambered for a Long Rifle cartridge will eventually cause some chamber erosion in front of where the Short case ends. When a Long or Long Rifle cartridge is fired in a chamber so eroded, extraction of the fired case can be very difficult. Therefore, extensive use of any Short cartridge is not recommended if the firearm is also to be used with Long or Long Rifle ammunition.

In a firearm chambered and rifled for the Short only, the .22 Shorts are amazingly accurate and without recoil.

For use on small game at ranges to 25 or 30 yards, the .22

Short High-Speed Hollow Point will work well. Even in this configuration, the Short should never be used on anything larger than squirrels or rabbits, and then only at the ranges indicated.

In order to ensure good accuracy, the Short is most efficiently used in guns specifically chambered for Short cartridges only. The rifling twist used in Long Rifle firearms is not ideal for the Short round.

.22 LONG

The Long round persists in hanging on despite there being little reason for its survival. At one time, the Long was discontinued by one of the major manufacturers. However, the shooting public's demand was heard and the round was reinstated.

The Long is a cross between the Short and the Long Rifle cartridge. It uses the 29-grain bullet of the Short and the case of the Long Rifle. The muzzle velocity of the Long in the high-speed load (the only load currently produced) is 1240 fps. However, due to the light bullet weight, the energy level is substantially below the Long Rifle.

Applications of the Long are identical to the Short. Its extra 22 ft.-lbs. does not qualify it for anything over the Short's domain.

I am sure that eventually the Long cartridge will disappear, but I'm not taking any bets on when it will happen.

A .22 CB Long is also available. This is an ideal indoor cartridge when an adequate backstop is used. It is also excellent for backyard shooting where it is appropriate. It has one big advantage over the CB Short in that it can be manually functioned through almost any gun action. The CB Long uses a 29-grain bullet at approximately 675 fps.

.22 LONG RIFLE

Undoubtedly, the .22 Long Rifle is the most popular cartridge ever designed, and it will assuredly survive as long as we have the right to own and use firearms for recreation. There are a number of variations of this cartridge. Those currently being made are as follows:

> .22 Long Rifle Standard Velocity
> .22 Long Rifle High Velocity
> .22 Long Rifle High Velocity Hollow Point
> .22 Long Rifle Match Grade

The Winchester Xpediter and the CCI Stinger could also be

America's number-one rifle quarry—the grey squirrel. The best selections here are .22 Long Rifle high-speed hollow points and head shots.

considered variations of the Long Rifle. However, these cartridges actually use a case longer than the Long Rifle and could, therefore, be more correctly identified as "extra long rifle" cartridges. Their worth is questionable. Accuracy is not as good as it is with a Long Rifle cartridge, and shortly past 50 yards their ballistic level falls below the Long Rifle. This author views these cartridges more as marketing gimmicks than as noteworthy additions to the Long Rifle variation.

All the .22 Long Rifle cartridges are suitable for all small game to 75 or 80 yards.

The high-speed hollow-point variations can be used to 50 or 60 yards on varmint as large as woodchucks, if shooting is restricted to head shots. Body shots should never be taken on animals this tough or large, regardless of how short the range. To do so would ensure lost cripples.

For squirrel hunting, head shots are preferred to prevent damaging too much meat. Some squirrel hunters will not use hollow points for the same reason.

Equipped with good sights, most of the better-grade .22's will shoot groups between 1 to 1½ inches at 50 yards with Long Rifle ammunition.

Fine-grade, competitive small-bore shooters using match-grade ammo know full well that 1-inch groups, with their equipment, are possible at ranges of 100 yards.

The Long Rifle cartridge is a favorite for just about any small game. Varmints as large as fox and as tough as woodchucks can be successfully hunted with it. However, it is a hunter's cartridge. It teaches you to stalk game carefully to ensure the swift and clean bagging of your game, and that is really half the fun of hunting with a .22 rifle. The low cost of ammo provides almost anyone with the opportunity of becoming an expert with the cartridge. Plinking "matches" are often enjoyed at a full 100 yards with most shooters doing very well indeed.

Many a long-range varmint *shooter*, who has lost the thrill of the hunt, finds himself returning to the *hunting* of varmints with the .22 Long Rifle. With its low noise level, hunters find that they are welcomed in places that centerfire rifle shooters find closed to them.

.22 WINCHESTER MAGNUM RIMFIRE

With more energy at 100 yards than any Long Rifle cartridge has at the muzzle, the .22 Winchester Magnum Rimfire is the muscle cartridge of the .22 rimfires. Useful for all the game purposes as the Long Rifle, the .22 Mag. can stretch the range of varmint hunters to 100 yards and, with good equipment, possibly to 125 yards for woodchucks, etc. Despite ballistic levels at ranges beyond this, the Mag. is not practical due to accuracy limitations.

However, this increased range doesn't come without a premium. The cost of ammo is considerably more than the Long Rifle. While it is still far cheaper than any centerfire cartridge, the cost is a bit high for plinking. It is a cartridge for the serious rimfire hunter who is willing to purchase equipment that will

The favorite varmint target—the woodchuck. With high-speed hollow points, the .22 Long Rifle can be effective to 50 yards if only head shots are taken.

enable him to take advantage of the higher energy levels at the longer ranges.

The noise level of the .22 Winchester Magnum Rimfire is noticeably above the Long Rifle, but it does not approach being objectionable except, perhaps, in very densely populated areas. There is a hollow-point version for most hunting and a full metal case version for situations where minimum pelt destruction or meat destruction is of importance.

5mm REMINGTON RIMFIRE MAGNUM

The muzzle velocity of 2105 fps makes this .20-caliber rifle the fastest rimfire round being produced today. It has a muzzle energy 50 ft.-lbs. greater than the .22 Winchester Magnum Rimfire, and at 100 yards it has 62 ft.-lbs. more energy.

With all of this going for it, one would expect the little 5mm to be a very popular cartridge. Not so. The cartridge is already dying after only a brief period in the marketplace. It never had a chance to develop a reputation of any kind. It may have been simply that the public just did not want a cartridge of smaller caliber than the .22. It may have been that none of the Remington guns chambered for it were popular, or perhaps the rumored problems with the cartridge were very real. In any event, ammo is already very difficult to find on most ammunition dealers' shelves.

No shooter should be without a .22 caliber rimfire firearm. Inexpensive practice, the fun of plinking, serious hunting, or target shooting can all be enjoyed almost anywhere with a .22. It truly is the most valuable recreational rifle or pistol cartridge ever designed.

CHAPTER TWO

Rimfire Exterior Ballistics Tables for Rifles and Handguns

The reader should keep in mind, when using the ballistic tables in this chapter, that all the data is basically computer developed, and is based on the nominal averages of ammunition.

The velocities listed are not necessarily the velocities you will obtain in your particular rifle or with your particular lot of ammunition. Average velocities from one firearm to the next will vary, due to manufacturing tolerances in the firearms. Also, one lot of ammunition may very well deliver a different average velocity than another, again due to manufacturing tolerances.

The reader should further be aware of the fact that average velocities are simply the medium of a range of velocities obtained over a test sample. It would be common in a string of ten shots of rimfire ammunition to have the highest and lowest velocities recorded as much as 50 fps or more apart.

Therefore, when reviewing, say, the maximum range (or any other table) figures, one should keep in mind that such range is determined by an *average* velocity, an *average* bullet weight, and an *average* bore diameter, etc. None of the figures in a ballistics table can be considered an absolute. Rather they should be considered as indications of approximately what can be expected as an average result.

Exterior ballistics of rimfire cartridges are affected by the barrel length of the firearm. However, with rimfire cartridges, the change in velocity per inch of barrel is rather modest. All the rifle exterior ballistics tables in this chapter have been developed with a barrel length of 24 inches. The only exceptions are the .22 Winchester Auto. ballistics which were developed with a 20-inch barrel.

The rimfire rifle ballistics tables include data for hold under

CASE OPERATIONS

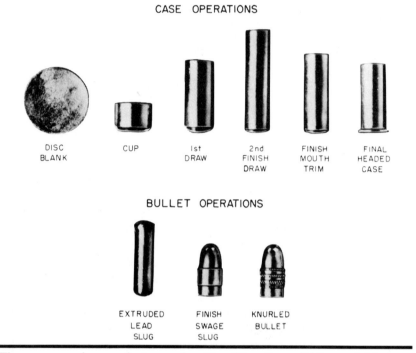

| DISC BLANK | CUP | 1st DRAW | 2nd FINISH DRAW | FINISH MOUTH TRIM | FINAL HEADED CASE |

BULLET OPERATIONS

| EXTRUDED LEAD SLUG | FINISH SWAGE SLUG | KNURLED BULLET |

The process of manufacture of a .22-caliber Long Rifle cartridge.

for a 45-degree hill at 100 yards. This data may be computed to any other angle by using the following formula:

$$h \left(\frac{N}{45} \right)^2 = \text{hold under for new angle in which:}$$

h = hold under for a 45-degree angle. Obtain this number from ballistics tables for the exact load you are using.

N = new angle for which you wish to calculate hold under.

For example:

What is the hold under for the .22 Long Rifle hollow point at 1280 fps muzzle velocity at 100 yards for an angle of 30 degrees? Referring to line 10 of the appropriate ballistics table we find that the hold under for this load at 100 yards and a 45-degree angle is 3.8 inches. Therefore:

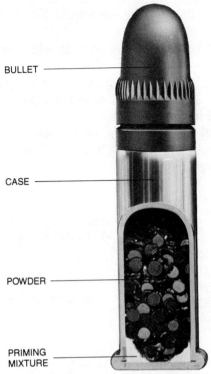

BULLET

CASE

POWDER

PRIMING
MIXTURE

Cutaway view of a completed .22 Long Rifle round.

$$h \left(\frac{N}{45} \right)^2 = \text{hold under}$$

$$h \left(\frac{30}{45} \right)^2 = \text{hold under}$$

$$3.8 \left(\frac{30}{45} \right)^2 = \text{hold under}$$

$3.8 \, (.6666)^2 = \text{hold under}$
$3.8 \times .4444 = \text{hold under}$
$1.68'' = \text{hold under for new angle of 30 degrees}$

Keep in mind that the hold under distance you arrive at from either the ballistics tables or your calculations does not mean that you would aim that distance below your target. Rather it means that you would subtract the distance arrived at from the required normal hold over.

In the instance described above, the normal hold over for a gun sighted-in at 50 yards would be 6.1 inches. We, therefore, subtract the hold under of 1.68 inches from the normal hold over (on flat land) of 6.1 inches. Our new hold over for a 30-degree angle is then reduced to 4.42 inches.

You may also calculate the height of trajectory when your sight height differs from the tables 0.9 inch above bore line. To do so, use the following formula:

$$\left[\left(\frac{R_t}{R_s}\right) - 1\right](h - 0.9) = \text{new height of trajectory to be added to original height of}$$

trajectory where:
R_t = target range in yards
R_s = sight-in range in yards
h = new neight of sights above bore line in inches

For example:

What would the new height of trajectory be for a .22 Long Rifle hollow point (37 grains) at a velocity of 1280 fps when the sights are 2 inches above bore line, the distance is 100 yards, and the zero was at 50 yards?

Refer to the proper ballistics table on line 10 under "height of trajectory (inches) above line of sight." Do not use "midrange trajectory height inches from bore line." You will find that the original trajectory was −6.1 inches. We then have:

$$\left[\left(\frac{R_t}{R_s}\right) - 1\right](h - 0.9) = \text{new height of trajectory to be added to original } -6.1 \text{ inches}$$

$$\left[\left(\frac{100}{50}\right) - 1\right](2.0 - 0.9) = \text{new height of trajectory to be added to original } -6.1 \text{ inches}$$

$[(2) - 1](1.1)$ = new height of trajectory to be added to original −6.1 inches
1×1.1 = new height of trajectory to be added to original −6.1 inches
−5.0 inches = new height of trajectory

Due to the large amount of data listed for each specific load, it was impossible to list all the data for a given cartridge on one page. Therefore, the use of the tables involves a line code.

To use the tables, simply locate the cartridge and load of your choice on the first table. Note that a line number is listed to the right of each cartridge and load. Simply refer to that line number on the following tables to locate the ballistics that apply to your load.

The conditions that apply to the ballistics tables are:

Temperature = 59.0° F.
Barometer = 29.53″ of mercury
Elevation = 0.0 feet

EXTERIOR BALLISTICS TABLES—RIMFIRE RIFLE

Line No.	Cartridge	Wgt. Grs.	Ballistic Coefficient	Type	0	100 Yards	200	0	100 Yards	200
			Bullet			Velocity in fps			Energy in ft.-lbs.	
1	.22 Short	15	0.043	Dis.	1690	928	668	95	29	15
2	.22 Short H.V.	15	0.043	Dis.	1790	951	605	107	30	16
3	.22 Short	29	0.083	Dis.	1045	872	736	70	49	35
4	.22 Short S.V.	29	0.083	Ball	1045	872	736	70	49	35
5	.22 Short H.V.	27	0.078	H.P.	1120	904	753	75	29	34
6	.22 Short H.V.	29	0.083	Ball	1095	902	761	77	52	37
7	.22 Long H.V.	29	0.083	Ball	1240	961	809	99	59	42
8	.22 Long Rifle S.V.	40	0.115	Ball	1150	975	859	117	84	66
9	.22 Long Rifle H.V.	36	0.103	H.P.	1280	1009	875	131	81	61
10	.22 Long Rifle H.V.	37	0.106	H.P.	1280	1013	882	135	84	64
11	.22 Long Rifle H.V.	38	0.109	H.P.	1280	1017	888	138	87	67
12	.22 Long Rifle H.V.	40	0.115	Ball	1255	1016	893	140	92	71
13	.22 Win. Auto.	45	0.126	Ball	1035	917	819	107	84	67
14	.22 Win. R.F. H.V.	45	0.126	Ball	1320	1054	931	174	111	87
15	.22 Win. Mag. R.F.	40	0.110	FMC, JHP	1910	1326	1008	324	156	90
16	5mm Rem. R.F. Mag.	38	0.146	JHP	2105	1609	1229	374	218	127
17	.25 Stevens Short	65	0.140	Ball	935	844	763	126	103	84
18	.25 Stevens Long	65	0.140	Ball	1115	981	884	179	139	113
19	.32 Short	80	0.120	Ball	935	830	738	155	122	97
20	.32 Long	80	0.120	Ball	1030	908	806	188	146	115

H.V.—high velocity
S.V.—standard velocity
R.F.—rimfire
Dis.—disintegrating
H.P.—hollow point
FMC—full metal case
JHP—jacketed hollow point

EXTERIOR BALLISTICS TABLES—RIMFIRE RIFLE

Line No.	Cartridge	Time of Flight Seconds 100 Yards	Time of Flight Seconds 200 Yards	Bullet Drop Inches from Bore Line 100 Yards	Midrange Trajectory in Inches from Bore Line 100 Yards	Drift-Ins 10 mph Wind 100 Yards	45° Angle Up- or Downhill Hold (Inches) Under Target 100 Yards	Height of Trajectory (Inches) Over Sight Line for Sights 0.9 Inches Over Bore Line 50 Yards	100 Yards	150 Yards
1	.22 Short	0.257	0.640	10.66	3.3	14.0	3.1	0	-5.7	-21.6
								2.9	0	-13.0
2	.22 Short H.V.	0.246	0.620	9.65	3.0	13.8	2.8	0	-5.1	-19.9
								2.6	0	-12.2
3	.22 Short	0.316	0.691	18.13	4.8	5.0	5.3	0	-8.7	-28.9
								4.4	0	-15.8
4	.22 Short S.V.	0.316	0.691	18.13	4.8	5.0	5.3	0	-8.7	-28.9
								4.4	0	-15.8
5	.22 Short H.V.	0.301	0.666	16.38	4.4	5.9	4.8	0	-7.9	-26.5
								3.9	0	-14.6
6	.22 Short H.V.	0.304	0.667	16.78	4.5	5.3	4.9	0	-8.0	-26.8
								4.0	0	-14.7
7	.22 Long H.V.	0.281	0.622	14.14	3.9	6.9	4.1	0	-6.8	-23.1
								3.4	0	-12.8
8	.22 Long Rifle S.V.	0.286	0.614	14.99	4.0	4.4	4.4	0	-7.0	-23.2
								3.5	0	-12.7
9	.22 Long Rifle H.V.	0.270	0.590	13.03	3.5	6.2	3.8	0	-6.2	-20.9
								3.1	0	-11.6
10	.22 Long Rifle H.V.	0.269	0.587	12.98	3.5	6.1	3.8	0	-6.1	-20.8
								3.1	0	-11.5
11	.22 Long Rifle H.V.	0.268	0.585	12.92	3.5	6.0	3.8	0	-6.1	-20.6
								3.1	0	-11.4
12	.22 Long Rifle H.V.	0.270	0.586	13.21	3.6	5.5	3.9	0	-6.2	-20.8
								3.1	0	-11.5
13	.22 Win. Auto.	0.307	0.655	17.67	4.6	3.3	5.2	0	-8.3	-26.9
								4.1	0	-14.5
14	.22 Win. R.F. H.V.	0.259	0.562	12.01	3.3	5.5	3.5	0	-5.6	-19.0
								2.8	0	-10.6
15	.22 Win. Mag. R.F.	0.190	0.454	6.16	1.7	5.7	1.8	0	-2.6	-10.1
								1.3	0	-6.3
16	5mm Rem. R.F. Mag.	0.163	0.378	4.71	1.3	3.6	1.4	0	-1.7	-6.7

								0.8	0	−4.2
17	.25 Stevens Short	0.338	0.712	21.31	5.5	3.0	6.2	0	−10.1	−32.4
								5.1	0	−17.3
18	.25 Stevens Long	0.288	0.611	15.41	4.0	3.4	4.5	0	−7.2	−23.3
								3.6	0	−12.6
19	.32 Short	0.341	0.725	21.57	5.6	3.5	6.3	0	−10.3	−33.3
								5.2	0	−17.8
20	.32 Long	0.311	0.662	17.90	4.7	3.5	5.2	0	−8.4	−27.4
								4.2	0	−14.8

EXTERIOR BALLISTICS TABLES—RIMFIRE RIFLE

Bullet

Line No.	*Cartridge*	*Wgt. Grs.*	*Ballistic Coefficient*	*Type*	*Muzzle Velocity (fps)*	*Maximum Range (yds.)*	*Velocity at Max. Range (fps)*	*Angle of Elevation (deg.)*
1	.22 Short	15	0.043	Dis.	1690	803	153	27.39
2	.22 Short H.V.	15	0.043	Dis.	1790	811	153	27.25
3	.22 Short	29	0.083	Dis.	1045	1210	206	31.16
4	.22 Short S.V.	29	0.083	Ball	1045	1210	206	31.16
5	.22 Short H.V.	27	0.078	H.P.	1120	1175	200	30.58
6	.22 Short H.V.	29	0.083	Ball	1095	1232	207	30.95
7	.22 Long H.V.	29	0.083	Ball	1240	1271	208	30.57
8	.22 Long Rifle S.V.	40	0.115	Ball	1150	1589	240	32.10
9	.22 Long Rifle H.V.	36	0.103	H.P.	1280	1506	230	31.37
10	.22 Long Rifle H.V.	37	0.106	H.P.	1280	1538	233	31.49
11	.22 Long Rifle H.V.	38	0.109	H.P.	1280	1568	236	31.60
12	.22 Long Rifle H.V.	40	0.115	Ball	1255	1623	241	31.86
13	.22 Win. Auto.	45	0.126	Ball	1035	1632	249	32.93
14	.22 Win. R.F. H.V.	45	0.126	Ball	1320	1757	253	32.13
15	.22 Win. Mag. R.F.	40	0.110	FMC, JHP	1910	1715	237	30.37
16	5mm Rem. R.F. Mag.	38	0.146	JHP	2105	2176	275	31.48
17	.25 Stevens Short	65	0.140	Ball	935	1659	258	33.99
18	.25 Stevens Long	65	0.140	Ball	1115	1819	263	33.06
19	.32 Short	80	0.120	Ball	935	1486	241	33.32
20	.32 Long	80	0.120	Ball	1030	1569	243	32.74

EXTERIOR BALLISTICS TABLE—RIMFIRE HANDGUNS
Bullet

Cartridge	Wgt. Grs.	Type	Barrel Length	Muzzle Velocity (fps)	Muzzle Energy (ft.-lbs.)
.22 Short H.V.	29	all	6″	1010	66
.22 Short S.V.	29	all	6″	865	48
.22 Long H.V.	29	all	6″	1095	77
.22 Long Rifle H.V.	40	all	6″	1060	100
.22 Long Rifle S.V.	40	all	6″	950	80
.22 Long Rifle Match (Pistol)	40	all	6.75″	1060	100
.22 Win. Mag. R.F.	40	all	6.5″	1480	195

CHAPTER THREE

Recoil Calculation, an Important Step in Cartridge or Shell Selection

Before getting into the specifics of centerfire cartridge or shell selection one should have a good understanding of recoil. Recoil in firearms has long been one of the most important criteria used by shooters in the selection of a cartridge or a shotshell gauge. However, there is a lot of misunderstanding on how much a given cartridge or gauge will recoil. For instance, some 20-gauge loads fired in the average weight 20-gauge gun will have more recoil than some 12-gauge loads fired in the average weight 12-gauge gun.

Many a shooter has had his or her ability to shoot seriously hampered by the selection of a cartridge or load that simply had too much recoil for him or her to handle. Knowing how to calculate the recoil of any cartridge/gun combination can help you to make a sensible selection of the cartridge/gun combination.

Little is gained by being so handicapped by recoil that a painful experience can cause the shooter to lose interest or, worse yet, develop a flinch.

A new shooter should carefully note the recoil levels of various combinations and work his or her way up the recoil ladder, staying with each load until there is no noticeable sensation of recoil. Doing so will prevent any gunshyness or flinching; and surely the development of a far more skilled shooter results when one pays attention to free recoil energy.

When reviewing a carefully prepared chart one can make intelligent choices on specific loads. Keep in mind, however, that certain other factors are important with respect to the actual sensation of perceived recoil. What you actually feel with regard to recoil depends upon how the recoil is delivered to, and absorbed by, your body. A poorly fitting gun surely seems to kick harder than a well-fitting gun (all other factors being equal). The sensation of recoil is greatly influenced by stock design, butt plate design, how the gun is held, and so on. One should always take advantage of the use of a recoil pad on all but the lightest-kicking guns or perhaps on special-purpose guns. Additionally, the nonslip surface of a recoil pad will protect your gun from bad bangs as a result of a gun slipping after it has been standing in a corner.

Semiautomatic guns deliver recoil that is perceived by the shooter as being substantially less than that delivered by a fixed-breech gun. This fact explains, at least in part, the great popularity of semiautomatic guns.

On a fixed-breech shotgun system, normal ignition/barrel time for acceptable loads is in the order of 3 milliseconds, with 3.5 milliseconds being about maximum. This is the total time in which the originally motionless gun is acted upon by the forces that generate recoil. In a semiautomatic shotgun, the same ignition/barrel time occurs. However, we have another factor coming into play: the moving parts of the gun mechanism. These include all the various parts of the gas systems, bolts, springs, and so forth. Their movement has the effect of storing the recoil energy and spreading out the time of application of recoil to the shooter. The cycle time of most semiautomatic shotguns is in the order of 10 to 15 milliseconds, depending upon gun characteristics, model, and so on. This spreads our perceived or felt recoil over a longer period for semiautomatics than for the fixed-breech guns. This time spread is enough to cause the sensation of recoil to appear to be milder. It is much like someone applying a sharp blow to your shoulder in comparison to someone pushing on your shoulder more slowly yet applying the same total force.

When properly instructed in safe gun handling, a new shooter will always find a semiautomatic more enjoyable and easier to shoot, all other things being equal. The same basics apply to semiautomatic rifles and handguns.

Calculating free recoil energy is not at all difficult. Armed with the basic equation, one need only insert the proper values and quickly he or she will have the recoil value of any specific combination. One of the laws of physics states that for every action there is an opposite and equal reaction. In a shotgun, rifle, or handgun the action that is generated by the burning gases pushing the ejecta from the muzzle is, of course, creating a reaction. This reaction is what we term recoil.

Recoil calculations are based on the gun weight, the weight of the wad (if any), the weight of the shot charge or bullet, the muzzle velocity of the shot charge or bullet, the weight of the powder, and the velocity of the escaping gases at the muzzle.

Technically speaking, recoil is based on the law of the conservation of momentum. Some of you may remember from your physics class that this law states: If a force and its reaction act between two bodies (with no other forces present), equal and opposite changes in momentum will be given to the two bodies. The first body, of course, is the weight of the ejecta driven from the muzzle. (The weight of the ejecta includes the combined weights of the wad and the shot charge or bullet.) The second body is, of course, your gun.

So, the momentum of a free recoiling gun is equal and opposite in direction to the momentum of the ejecta.

Mathematically expressed we have:

$$\text{Wgt.}_{\text{gun}} \times \text{Vel.}_{\text{gun}} = \text{Wgt.}_{\text{ejecta}} \times \text{Vel.}_{\text{ejecta}} + \text{Wgt.}_{\text{powder}} \times \text{Vel.}_{\text{powder gases}}$$

or simply: $W_g V_g = W_e V_e + W_p V_p$

All weights are in pounds and all velocities in fps.

Turning this about, the velocity of free recoil of a firearm is, therefore, as follows:

$$\text{Vel.}_{\text{gun}} = \frac{\text{Wgt.}_{\text{ejecta}} \times \text{Vel.}_{\text{ejecta}} + \text{Wgt.}_{\text{powder}} \times \text{Vel.}_{\text{powder gases}}}{\text{Wgt.}_{\text{gun}}}$$

or: $V_g = \dfrac{W_e V_e + W_p V_p}{W_g}$

Naturally all our weights are readily determined with the use of appropriate scales, and by chronograph we determine the velocity of the ejecta.

The effective velocity of the powder gases has been found, for all practical purposes, to be equal to the velocity of the ejecta multiplied by a factor of 1.5. This relationship has been established by well-documented scientific experiments both in Great

Britain and the United States. Hence, our formula for velocity of free recoil of a gun becomes:

$$V_g = \frac{W_e V_e + 1.5 W_p V_e}{W_g}$$

To calculate the energy of free recoil, we have:

Kenetic energy = ½ Mass × Velocity²

or K.E. = ½MV²

in which M = Mass = $\dfrac{\text{weight in lbs.}}{32.174 \text{ (gravitational constant)}}$

and V = velocity.

Therefore, K.E. or recoil can be expressed by:

Free Recoil Energy = $\dfrac{W_g V_g{}^2}{64.348}$

By substitution of $V_g{}^2$ we finally arrive at the standard equation:

Free Recoil Energy = $\dfrac{(W_e V_e + 1.5 W_p V_e)^2}{64.348 \, W_g}$

Wherein W_e = Weight of ejecta (shot and wad or bullet in lbs.). To obtain weight in pounds divide grain weights by 7000. To convert ounces to pounds divide by 16.

V_e = Velocity of ejecta in feet per second. Obtain this speed from ballistic tables in this book (under muzzle velocity).

W_p = Weight of powder in lbs. To obtain weight in pounds divide grain weight by 7000.

W_g = Weight of gun in pounds.

Assuming a 7½-pound shotgun, 1200 fps, a shot weight of 1⅛ ounces, a wad weight of 38.0 grains, and a powder charge of 20.5 grains, we would have a free recoil of 19.2 foot-pounds.

The foregoing formula can be rewritten as:

Free Recoil Energy = $\dfrac{[V_e(W_e + 1.5 W_p)]^2}{64.348 \, W_g}$

This last formula is easier to use long hand or on an inexpensive calculator than the so-called standard formula. It simply is the result of applying the distributive principle of multiplication over addition.

Do not round values when working through the formula. Early rounding can create substantial errors.

As you can see, calculating the recoil for any specific load and gun combination is quick and easy. Powder-charge weights can be obtained by breaking down a factory load or from a handloading data book.

IMPORTANT: When working with a load that has a muzzle velocity in excess of 2499 fps, the constant 1.5 in the formula should be changed to a constant of 1.75.

The following charts of shotshell/shotgun combinations were developed using the foregoing method. They will prove most useful in allowing you to select suitable loads for new shooters, for advancing shooters through various recoil levels, for selecting loads for shooters who are recoil shy, and so on. It should be kept in mind that a shooter can usually handle far more recoil when hunting as opposed to target shooting. Obviously, the exact recoil of a load can vary slightly from brand to brand due to the variables of powder-charge weight and wad weight. However, the charts will give you a relative value that will be very close to the actual value of your selected load. Velocities shown are industry nominals.

12-GAUGE, 3″ CHAMBER, 8 LBS. GUN WEIGHT

Shot Charge Weight	Velocity		Free Recoil Energy
1⅞ ounces	4 dram equivalent, 1210 fps	=	47.7 ft.-lbs.
1⅝ ounces	4 dram equivalent, 1280 fps	=	41.5 ft.-lbs.
1⅜ ounces	3¾ dram equivalent, 1295 fps	=	31.8 ft.-lbs.

12-GAUGE, 2¾″ CHAMBER, 7½ LBS. GUN WEIGHT

Shot Charge Weight	Velocity		Free Recoil Energy
1½ ounces	3¾ dram equivalent, 1260 fps	=	36.7 ft.-lbs.
1¼ ounces	3¾ dram equivalent, 1330 fps	=	29.7 ft.-lbs.
1¼ ounces	3¼ dram equivalent, 1220 fps	=	24.0 ft.-lbs.
1⅛ ounces	3½ dram equivalent, 1330 fps	=	24.5 ft.-lbs.
1⅛ ounces	3¼ dram equivalent, 1255 fps	=	21.4 ft.-lbs.
1⅛ ounces	3 dram equivalent, 1200 fps	=	19.2 ft.-lbs.
1⅛ ounces	2¾ dram equivalent, 1145 fps	=	17.4 ft.-lbs.
1 ounce	3¼ dram equivalent, 1290 fps	=	18.2 ft.-lbs.

16-GAUGE, 2¾" CHAMBER, 7 LBS. GUN WEIGHT

Shot Charge Weight	Velocity		Free Recoil Energy
1¼ ounces	3¼ dram equivalent, 1260 fps	=	27.9 ft.-lbs.
1⅛ ounces	3¼ dram equivalent, 1295 fps	=	24.3 ft.-lbs.
1⅛ ounces	2¾ dram equivalent, 1185 fps	=	20.1 ft.-lbs.
1 ounce	2½ dram equivalent, 1165 fps	=	15.0 ft.-lbs.

20-GAUGE, 3" CHAMBER, 7 LBS. GUN WEIGHT

Shot Charge Weight	Velocity		Free Recoil Energy
1¼ ounces	3 dram equivalent, 1185 fps	=	23.7 ft.-lbs.

20-GAUGE, 2¾" CHAMBER, 6¾ LBS. GUN WEIGHT

Shot Charge Weight	Velocity		Free Recoil Energy
1⅛ ounces	2¾ dram equivalent, 1175 fps	=	19.9 ft.-lbs.
1 ounce	2¾ dram equivalent, 1220 fps	=	17.3 ft.-lbs.
1 ounce	2½ dram equivalent, 1165 fps	=	15.7 ft.-lbs.
⅞ ounce	2½ dram equivalent, 1210 fps	=	13.2 ft.-lbs.

28-GAUGE, 2¾" CHAMBER, 6½ LBS. GUN WEIGHT

Shot Charge Weight	Velocity		Free Recoil Energy
¾ ounce	2¼ dram equivalent, 1295 fps	=	12.1 ft.-lbs.

.410 BORE, 3" CHAMBER, 5½ LBS. GUN WEIGHT

Shot Charge Weight	Velocity		Free Recoil Energy
¹¹⁄₁₆ ounce	Maximum, 1135 fps	=	7.6 ft.-lbs.

.410 BORE, 2½" CHAMBER, 5½ LBS. GUN WEIGHT

Shot Charge Weight	Velocity		Free Recoil Energy
½ ounce	Maximum, 1135 fps	=	4.8 ft.-lbs.

The following rifle-cartridge recoil calculations are all based on a single gun weight for ease of comparison. Velocities are advertised nominals.

GUN WEIGHT 8 LBS.

Caliber	Bullet Weight	Velocity		Free Recoil Energy
.22 Hornet	45 grains	2690 fps	=	1.2 ft.-lbs.
.222 Remington	50 grains	3140 fps	=	3.3 ft.-lbs.
.22-250 Remington	55 grains	3730 fps	=	8.5 ft.-lbs.
.243 Winchester	80 grains	3420 fps	=	12.7 ft.-lbs.
.243 Winchester	100 grains	2960 fps	=	10.7 ft.-lbs.
.250-3000 Savage	87 grains	3030 fps	=	8.9 ft.-lbs.
.250-3000 Savage	100 grains	2820 fps	=	8.3 ft.-lbs.
.257 Roberts	117 grains	2650 fps	=	10.4 ft.-lbs.
.25-06 Remington	120 grains	3060 fps	=	16.8 ft.-lbs.
.264 Winchester Magnum	100 grains	3620 fps	=	25.7 ft.-lbs.
.264 Winchester Magnum	140 grains	3140 fps	=	24.5 ft.lbs.
.270 Winchester	100 grains	3480 fps	=	18.2 ft.-lbs.
.270 Winchester	130 grains	3110 fps	=	21.3 ft.-lbs.
7mm Mauser	175 grains	2470 fps	=	14.9 ft.-lbs.
7mm Remington Magnum	175 grains	2860 fps	=	25.4 ft.-lbs.
.30 M-1 Carbine	110 grains	1990 fps	=	2.8 ft.-lbs.
.30-30 Winchester	170 grains	2200 fps	=	9.1 ft.-lbs.
.300 Savage	180 grains	2350 fps	=	12.5 ft.-lbs.
.308 Winchester	150 grains	2820 fps	=	17.1 ft.-lbs.
.30-06 Springfield	180 grains	2700 fps	=	16.9 ft.-lbs.
.300 Winchester Magnum	180 grains	3000 fps	=	33.2 ft.-lbs.
.338 Winchester Magnum	250 grains	2660 fps	=	40.2 ft.-lbs.
.35 Remington	200 grains	2080 fps	=	11.4 ft.-lbs.
.375 H&H Magnum	270 grains	2530 fps	=	42.2 ft.-lbs.
.44 Remington Magnum	240 grains	1760 fps	=	9.4 ft.-lbs.
.444 Marlin	240 grains	2350 fps	=	21.1 ft.-lbs.
.45-70 Government	405 grains	1330 fps	=	13.3 ft.-lbs.
.458 Winchester Magnum	510 grains	2110 fps	=	68.4 ft.-lbs.

The following handgun-cartridge recoil calculations are all based on a single gun weight for ease of use. Velocities are advertised nominals.

GUN WEIGHT 40 OUNCES

Caliber	Bullet Weight	Velocity		Free Recoil Energy
9mm Luger	100 grains	1320 fps	=	2.6 ft.-lbs.
.38 Special	110 grains +P	1020 fps	=	1.9 ft.-lbs.
.38 Special	158 grains	755 fps	=	2.0 ft.-lbs.
.357 Magnum	110 grains	1295 fps	=	3.2 ft.-lbs.
.357 Magnum	158 grains	1235 fps	=	6.1 ft.-lbs.
.44 Magnum	240 grains	1350 fps	=	13.7 ft.-lbs.
.45 ACP	230 grains	810 fps	=	4.7 ft.-lbs.
.45 Colt	255 grains	860 fps	=	6.6 ft.-lbs.

One should keep in mind that all the foregoing data is based on advertised velocities, which are not always the velocities you will obtain. It is also important to remember that a gun's weight plays a very influential role on recoil. Increase the gun weight by a factor of two, and recoil is cut in half. Conversely, reducing the gun weight by half will increase recoil by a factor of two. Many of the rifle calibers listed are available only in guns somewhat lighter or heavier than our arbitrary 8 pounds. You will have to calculate the recoil for your own combination to have exact figures. Be sure to include all accessories in the weight of the gun (scope, mounts, etc.).

The final consideration in determining the selection of a cartridge after going through the recoil calculations is consideration of the muzzle blast. So-called magnum-type cartridges, due to the relatively large amounts of powder and high muzzle velocities, tend to create a very loud muzzle blast. This ear-shattering noise is best left to those who have had some experience in this area. Many a shooter has developed a serious flinch from the muzzle blast in the .357 Magnum cartridge with 110-grain bullets. Here recoil is only moderate, but the noise is deafening.

While current trends seem to be toward guns with more recoil and usually more muzzle blast, the average shooter would do well to go in the other direction. His ability to shoot well and, therefore, perform well on targets or in the field will be greatly enhanced. As a result, he will enjoy his sport far more than if he were the victim of too much recoil or noise.

The average shooter using a rifle can usually handle up to 15 foot-pounds of recoil without noticing the actual recoil. Most beginners can handle 12 to 13 foot-pounds without any problems developing. There are, of course, exceptions in both directions. An experienced shot usually will find that about 17.5 foot-pounds will be the maximum for comfortable shooting.

In shotgun games where gun and body are kept in motion, about 2 or 3 foot-pounds can be added to the foregoing numbers.

Those who suffer from various physical handicaps will find perhaps that the selection of a cartridge/gun combination with recoil of no more than 10 or 11 foot-pounds might enable them to enjoy a sport they may otherwise have to give up.

In handguns, a beginner should stick with loads under 2.5 foot-pounds, while the more experienced shooter can handle up to 6.5 foot-pounds. It takes a lot of determination to handle recoil heavier than 7 foot-pounds in a handgun.

Making the right choice will ensure years of good shooting, even if you have not purchased a status symbol!

CHAPTER FOUR

Choosing a Centerfire Rifle Cartridge

The selection of a centerfire rifle cartridge has become needlessly complex and difficult. The total number of available cartridges to choose from far exceeds the number needed to make a selection for any intended purpose.

Additionally, the myriad of published words on this topic has, for the most part, only added to the confusion. Many of those who have written on the subject were highly qualified to do so; but equally as many had a pitifully small amount of knowledge or experience on which to base their comments.

In this chapter, we will attempt to treat cartridge selection as simply as possible. We will handle each of the cartridges currently available and list, by bullet weight, the appropriate applications. By doing so, the reader, whether a novice, a confused researcher, or an experienced shooter, will be able to tell at a glance for what purpose(s) a specific caliber and bullet weight are suited.

There is a multitude of overlapping choices available. The best selection becomes a matter of personal preference. In many cases, the type of firearm you prefer to use will narrow down the selection. For example, if you prefer a lever-action rifle for deer hunting you cannot choose the 7mm (7×57mm) Mauser cartridge since no lever guns are available in this caliber, and so on.

Each cartridge will be treated in ascending caliber designation. The reader wishing to select a deer cartridge will merely have to pick out those calibers listed for light big game and the appropriate bullet weight. The general comments may help to further shorten the list of choices available. If not, perhaps the comments will just add to your general knowledge of ammunition.

Everything from varmint to elephant will be included in the

selection guide. For ease in listing the cartridge/bullet weight application, the following categories have been used:

N.R.
Not recommended for any specific use. Refer to the comments under cartridges so marked for further clarification.
Varmint
Includes all such critters from ground squirrels through woodchucks, foxes, and on up to animals the size of coyotes.
Light Big Game
Includes animals the size of deer, antelope, and black bear, plus all others up to 500 pounds in weight.
Big Game
Includes all big-game animals in excess of 500 pounds that are *not* generally considered a threat to the life of the hunter.
Dangerous Game
Includes the larger bears and the African animals normally capable of inflicting serious harm upon the hunter.

Specific references will be made to range limitations of the various cartridges where appropriate. In general, short range means 50 yards or less, medium range means 150 yards or less, and long range means 300 yards or less.

Comments on the cartridge's popularity are made to help prevent the reader from selecting a caliber for which ammunition may be very difficult to find.

Some of the author's comments will undoubtedly bring cries of anguish from those who, for personal reasons, feel a cartridge is suited for more or less than indicated. The author simply suggests that he has heard all these arguments and stands firmly behind the following selection guide.

No attempt will seriously be made to trace cartridge history or development nor to explain the reasons for each cartridge designation. These subjects could fill an entire volume.

Our discussion will be limited to those cartridges currently being manufactured by Winchester-Wetern, Remington-Peters, Federal, and I.V.I. of Canada.

CENTERFIRE CARTRIDGE SELECTOR GUIDE

.17 CALIBER

.17 Remington
25-Grain Bullet—Varmint

While this is a useful varmint cartridge to about 225 yards, it tends to be better suited for those who are a bit more enthusiastic about specialty firearms than most shooters. Barrel-cleaning problems can prove to be nettlesome because of the bullet-jacket fouling that seems to be unique to this caliber. The chore is further complicated by the unavailability of proper diameter cleaning rods and barrel brushes. At best, this cartridge has not proved to be extremely popular.

.22 CALIBER

.218 Bee
46-Grain Bullet—Varmint

This varmint cartridge was introduced in the late 1930's. It is useful to about 150 yards in a bolt-action gun. Practical ranges for lever-action guns are greatly reduced. This cartridge is all but dead, and little reason exists to lament its passing as long as the .22 Hornet is with us.

.22 HORNET

45-Grain Bullet—Varmint
46-Grain Bullet—Varmint

This very accurate varmint cartridge is useful to about 125 yards. The Hornet ranks among the all-time-great cartridges with respect to accuracy. It would be difficult to find a factory gun/ammunition combination in this caliber that did not prove to be very accurate. Groups of less than 1 inch are commonplace with the .22 Hornet. This cartridge was being produced by the ammunition manufacturers long before commercial rifles were chambered for it.

This cartridge was dying until very recently. The craze for ultra-high velocity and long-range shooting had all but driven it into obsolescence. No firearms (except the Savage O/U) were being chambered for it until Ruger introduced the Number 3 single shot.

There are a great many shooters who realize that hunting in suburbia requires a relatively quiet and highly accurate cartridge. The Hornet, therefore, is now very much alive and well.

For those who have become bored with varmint shooting, as opposed to varmint hunting, the .22 Hornet in the single-shot Ruger will provide a great deal of satisfaction while forcing the shooter to become a skilled hunter. Anywhere that stalks on varmint can be made successfully to 125 yards or less is considered .22 Hornet country.

Another nice plus with the Hornet is its very long barrel life. A good Hornet barrel seems to last forever (a result of the moderate velocity and low powder-charge weight).

The soft-point bullets are greatly favored over the hollow-point bullets. The soft points as a whole seem to shoot a bit more accurately.

.22 Savage
70-Grain Bullet—N.R.
This cartridge is currently available only as an import from Canada. It was developed in the late 1890's by Charles Newton for Savage, who in turn introduced it as a deer cartridge. The ballistic level doomed it to failure from the start. At best, it makes a fair short-range varmint gun due to its limited accuracy.

This cartridge was unique because it used a .228-inch diameter bullet instead of the .224-inch diameter bullet used in all other factory .22-caliber cartridges. The cartridge is dead and is best left that way.

.222 Remington
50-Grain Bullet—Varmint
The .222 is a superbly accurate cartridge that ranks among the all-time greats. It is an excellent varmint load to about 250 yards. The noise level is moderate. It is, perhaps, the single most popular varmint cartridge and justifiably so. Almost any good rifle/ammunition combination will prove to be very accurate and effective.

The cartridge is a favorite among benchrest shooters because of its inherent accuracy. Where ranges do not exceed 250 yards and noise of a moderate level is acceptable, it would be difficult to make a better selection.

.222 Remington Magnum
55-Grain Bullet—Varmint
There was a time—maybe it's still with us in part—that the word magnum sold a lot of guns and ammunition. After Remington's success with the .222 Remington, the natural next step was to magnumize the cartridge. Remington made the case a bit longer, added the word magnum, and sat back to wait for its newest success. Perhaps Remington is still waiting. In any event, the slightly heavier bullet and a bit more velocity were more than offset by the .222 Remington's superior accuracy. The magnum never did realize a great amount of popularity.

The .222 Remington Magnum brass case, however, is a favorite of benchrest shooters who use it as a basis to form many of their wildcat cartridges.

.223 Remington
55-Grain Bullet—Varmint
This is the commercial version of the current U.S. military cartridge (5.56mm). Ballistics are about on par with the .222 Remington Magnum. Due to a difference in throat configuration and loading specifications, use of military ammunition in sporting chambers can cause pressure increases of 5000 to 8000 psi (pounds per square inch). Several semi-automatic rifles are available for this cartridge, and this type of rifle is favored by the plinkers.

.225 Winchester
55-Grain Bullet—Varmint
This cartridge never really took hold. Because of its erratic performance, it has been both praised and damned for its accuracy and

ballistic uniformity. This sometimes hot, sometimes cold performance is undoubtedly responsible for the cartridge's lack of popularity. The .225 is dying very quickly.

.22-250 Remington
55-Grain Bullet—Varmint

This is the ultimate long-range .22-caliber cartridge. It has totally replaced the now obsolete .220 Swift ammunition (rifles are still chambered for the Swift). Its accuracy is superb, and barrel life is a great deal better than the old .220 Swift. The cartridge is extremely popular with long-range varmint hunters and is used frequently on varmints up to 400 yards.

The .22-250 started life as a wildcat cartridge and was with us a great number of years before Remington legitimized it as a factory loading.

Noise level is high, and the cartridge is not popular in populated areas.

.24 CALIBER (6mm)

.243 Winchester
80-Grain Bullet—Varmint
100-Grain Bullet—Light Big Game

The .243 Winchester is a necked-down version of the .308 case. It is extremely popular as an ultra-long-range varmint cartridge. It is often used at ranges in excess of 400 yards. Obviously, the accuracy of this cartridge is excellent. It enjoys the popularity of being the best of the ultra-long-range varmint rifles.

The .243 has also proved itself to be a very effective round with the 100-grain bullet on animals the size of antelope, deer, etc. It is probably the very best choice as a combination long-range varmint and deer-sized animal rifle. Its modest recoil lends to its popularity.

The 1-10-inch twist rate is probably in part responsible for the .243's popularity being greater than that of the 6mm Remington.

6mm Remington
80-Grain Bullet—Varmint
90-Grain Bullet—Light Big Game
100-Grain Bullet—Light Big Game

When originally introduced as the .244 Remington, this cartridge was combined with a 1-12-inch twist. As a result, bullets heavier than 90 grains were not practical. It never proved very popular. Remington tried to give it new life by switching to a 1-9½-inch twist and renaming it the 6mm Remington. Despite this effort, it has never become quite as popular as Winchester's .243. The slightly larger case capacity results in a very slight increase in velocity over the .243. The difference is not enough to be of any consequence.

The 100-grain loading should not be used in rifles marked .244 since the twist rate is inadequate to stabilize the longer bullet.

.25 CALIBER

.25-06 Remington
87-Grain Bullet—Varmint
100-Grain Bullet—Light Big Game
120-Grain Bullet—Light Big Game

The .25-06 was with us many years as a wildcat cartridge before Remington standardized it in a factory loading. It is a necked-down version of the .30-06 Springfield cartridge. It is well suited to the entire range of light big game with the 120-grain bullet as the correct choice for animals over 250 pounds.

It also performs well as a long-range varmint cartridge, sometimes being used on varmints at ranges up to 400 yards.

Due to the very large case capacity (over-bore), barrel life is somewhat shortened in comparison to other .25-caliber rifles. Noise level is quite high, and the muzzle blast prevents some shooters from doing their very best. Accuracy is usually good, when the shooter is able to handle the extra noise. Recoil, while moderate, really leaves this a less than ideal choice for most youngsters or for most women.

.25-20 Winchester
86-Grain Bullet—N.R.

The .25-20 Winchester served as a medium-range varmint cartridge in some bolt-action guns when the 60-grain high-velocity loading was available. Ranges were drastically shortened in the lever-action guns of this caliber due to the inferior accuracy of the combination. Today, only the 86-grain lead-bullet load is available. This load is suitable only for short-range varmints or small game. The cartridge has little overall value to recommend it for any purpose. It is a dead cartridge; no guns have been chambered for it for a long, long time. At best, the .25-20 is a curio left over from yesteryear.

.25-35 Winchester
117-Grain Bullet—Small Deer

During its years of semi popularity, the .25-35 gained a reputation of being a very accurate lever-action deer cartridge. It is highly unlikely that its so-called accuracy was the result of anything more than a very low level of recoil and noise, which consequently allowed the shooter to do his very best.

The ballistics of the .25-35 Winchester are marginal for deer hunting. It is useful only if great care is taken in placing good shots on standing game at short ranges.

The cartridge is all but dead, and no commercial firearms have been chambered for it for many years.

.250 Savage (.250-3000)
87-Grain Bullet—Varmint
100-Grain Bullet—Light Big Game

The .250 was originally introduced with the 87-grain bullet at a muzzle velocity of 300 fps. This cartridge was an instant success. The addition of the 100-grain bullet made it an ideal varmint/light big-game cartridge. Loaded to a more modest pressure level, the .250-3000 affords better barrel life than either of the 6mm's. While slightly lower velocity levels reduce effective range to something less than the 6mm's, it is still an excellent long-range cartridge (300 yards). Increased barrel life is one of its advantages over either of the 6mm's or the larger case .25's.

The .250's accuracy capabilities rank it among the all-time-great cartridges. Combine this with light recoil and a moderate noise level and you have the basis for this cartridge's popularity today. In fact, the .250 seems to be staging a justified comeback.

.256 Winchester
60-Grain Bullet—Varmint

This cartridge was originally introduced in the Ruger Hawkeye single-shot handgun. The world wasn't quite ready at that time for a single-shot varmint handgun, and the model was soon discontinued.

A few rifles were made for the cartridge, but it proved to be an extremely difficult cartridge to reload. As a result, the cartridge, which was a good 100-yard varmint load, is now almost forgotten.

.257 Roberts
87-Grain Bullet—Varmint
100-Grain Bullet—Light Big Game
117-Grain Bullet—Light Big Game

With its slightly larger case capacity and heavier bullet, the .257 Roberts is perhaps a bit more useful on light big-game animals between 300 and 400 yards than the .250 Savage. In all other ways, the .257 Roberts can be used for the same purposes and ranges as the .250-3000.

The 117-grain bullet should be selected for any game larger than deer and antelope in order to take advantage of the cartridge's potential.

While the .257 is a very accurate cartridge, the .250-3000, gun for gun, will usually outperform it. Where game larger than 300 pounds will normally be hunted, the .257 is to be preferred over the .250-3000. Recoil is light enough to make this a suitable choice for the shooter who prefers not to be aware of recoil.

.26 CALIBER (6.5mm)

6.5mm Mannlicher Schoenauer
160-Grain Bullet—Light Big Game

While no ammunition has been made for many years for this cartridge in this country, it was, at one time, quite popular. Much of its popularity stemmed from the romance of the cartridge created by one or more African hunters. It has been credited with being a satisfactory elephant cartridge.

While undoubtedly at least one man killed quite a few elephants with the little 6.5 and full metal case bullets, its ballistics are really not any better than that of the .30-30.

For stunt shooting, its very long bullet and the resulting high sectional density make it possible to obtain rather deep penetration with full metal jacketed bullets.

In the original Mannlicher Schoenauer carbine, this cartridge makes a delightful deer rifle to 200 yards. However, for all practical purposes, the cartridge is a dead item.

The very rapid twist rate of the firearms chambered for the 6.5 precludes the use of bullets lighter than approximately 138 to 140 grains.

6.5×55mm (6.5 Swedish)
160-Grain Bullet—Light Big Game
For many years this cartridge was the official service cartridge of the Swedish government. It is useful on game up to 500 pounds and at ranges not exceeding 200 yards.

Very few commercial guns have been chambered for this cartridge as it easily fits into the dead-cartridge category. Ammunition is now obsolete in this country.

6.5mm Remington Magnum
120-Grain Bullet—Light Big Game
Suitable for light big game (up to 500 pounds), this comparatively new cartridge is already dying. Originally it was also loaded with a 100-grain bullet which was suitable for varmint. However, the low popularity of this cartridge has reduced it to a single bullet-weight offering.

.264 Winchester Magnum
100-Grain Bullet—Varmint
140-Grain Bullet—Light Big Game, Big Game
This cartridge has proved to be the most popular of the 6.5's. However, at best, its popularity is only fair. Initially it seemed that it might become a very popular cartridge. However, the horrendous noise level, very short barrel life, and a noticeable amount of recoil proved to be undesirable for most shooters.

It is unquestionably an excellent long-range cartridge. It has been used to 500 yards on all-sized game from varmints to antelope.

It will serve well with the 140-grain bullet on animals up to 900 or 1000 pounds. However, it would prove inadequate for grizzly or brown bear.

The noise level is high enough to cause flinching in all but the most experienced shooters.

.27 CALIBER

.270 Winchester
100-Grain Bullet—Varmint
130-Grain Bullet—Light Big Game, Big Game
150-Grain Bullet—Light Big Game, Big Game
Based on the .30-06 Springfield case, this cartridge has proved to be extremely popular since the mid-1920's when it was introduced.

The 100-grain varmint bullet, in practice, does not prove to be as good a varmint load as it appears on paper. Any varmint load in .27 caliber or larger, at best, should be recognized for what it is—a means of using a big-game rifle for practice during the off-season. Accuracy capabilities with the 100-grain bullet limit varmint range to about 250 yards.

No such limitations apply to the 130-grain bullet for light big game or big game. This cartridge has been used successfully by skilled riflemen on big game to ranges of 400 yards. The 130-grain bullet is the preferred light big-game or big-game bullet except where extensive meat destruction is of real concern. In such cases, the 150-grain bullet can be selected for ranges up to 250 yards.

Recoil and noise level fall into the moderate range. While the .270 is not a cartridge for the shy shooter, it will continue to be a favorite for many, many years, especially for very long-range big-game hunting. For such shooting you can hit more easily with this cartridge than with either the .30-06 or .308, due to its somewhat flatter trajectory.

The .270 is an excellent choice for light big-game and big-game hunting. However, for game of 800 pounds or more, there are more effective cartridges in the .30-caliber-plus range.

.28 CALIBER (7mm)

7mm Mauser (7×57mm)
175-Grain Bullet—Light Big Game
This is an ideal cartridge for light big game (up to 500 pounds), especially when moderate recoil is of prime concern. Very few commercial rifles have been chambered for this cartridge. However, it continues to remain a reasonably popular cartridge. Cartridges loaded with lighter bullet weights are available as imports from Canada. However, the 175-grain bullet seems to be the best selection.

7mm Remington Magnum
125-Grain Bullet—Varmint
150-Grain Bullet—Light Big Game
175-Grain Bullet—Big Game
This cartridge was an instant success when introduced by Remington.

The varmint load, at best, affords a 200- to 225-yard off-season practice load. Cartridges this large cannot be truly satisfactory for varmint hunting.

The 150-grain bullet is ideal for all light big game, and the 175-grain bullet will prove effective on all but the biggest of our bears (brown and grizzly).

The cartridge's popularity attests its excellence in the field. It is not for those who find that noise and recoil reduce their shooting capabilities.

.280 Remington
150-Grain Bullet—Light Big Game
165-Grain Bullet—Big Game
Remington, I suspect, had hopes of introducing a cartridge that would compete with the .270 Winchester, when they introduced the .280. The best they had for their efforts was a less-than-popular cartridge. The cartridge has, despite its poor popularity, the capabilities of taking game from deer to moose.

.284 Winchester
125-Grain Bullet—Varmint
150-Grain Bullet—Light Big Game

Introduced in the short-action Winchester 88 and 100 Models, this cartridge was also another instant flop.

It is unique because it is the only American-manufactured rebated cartridge (the rim is considerably smaller than the case body).

No firearms are currently being chambered for the .284, and it can be considered a dead cartridge.

The 125-grain bullet in most firearms is a 225-yard cartridge. The 150-grain bullet is a 300- to 350-yard light big-game cartridge.

.30 CALIBER

.30 Carbine (30 M-1 Carbine)
110-Grain Bullet—N.R.

The carbine round is neither fish nor fowl. Too heavy for small game, too slow for varmint, and too light for light big game make for a cartridge for which there is no practical application.

Originally introduced in the military carbine, it was intended to replace a handgun for anti-personnel use. Even here it was woefully lacking. It is, however, a reasonably popular cartridge with plinkers. Sportsmanship should keep it from ever being used on deer, etc.

.30 Remington
170-Grain Bullet—Light Big Game

This cartridge is, ballistically speaking, a rimless equivalent of the .30-30. Scarcity of ammo, scarcity of guns, and no shooter interest relegate it to the dead-cartridge category.

.30-30 Winchester
125-Grain Bullet—Varmint
150-Grain Bullet—Light Big Game
170-Grain Bullet—Light Big Game

The .30-30 is one of the most popular cartridges ever produced. Undoubtedly, it has accounted for more deer than any other cartridge.

The 125-grain bullet, as loaded by Federal, is intended for short-range varmint shooting. It is effective to about 100 yards. The 150- and 170-grain bullets serve well on deer-sized animals to about 150 to 200 yards.

The .30-30 is an extremely accurate cartridge, which cannot be said of most of the guns so chambered. A few .30-30 bolt-action rifles have been produced, and these have always proved to be extremely accurate.

Because of modest recoil and noise, availability of ammo, and the light compact guns chambered for it, the .30-30 will continue to be our most popular deer cartridge for a long, long time. As long as shots are placed well and ranges kept below 200 yards, it will perform admirably on deer.

Most gun editors take Freudian delight in maligning the .30-30 as a barely adequate, inaccurate cartridge. However, this is simply not so. Properly handled, it will get the job done.

This is the progressive expansion of a .30-caliber, 180-grain power-point bullet.

.300 H&H Magnum
180-Grain Bullet—Big Game
220-Grain Bullet—Big Game

The .300 H&H Magnum is an excellent big-game cartridge and is suitable for long-range shooting. Its popularity, however, is declining fast since it has been almost entirely replaced by the .300 Winchester Magnum. The cartridge is extremely accurate and has accounted for wins at many 1000-yard matches.

The 180-grain bullet can be used on light big game, but one would be needlessly overgunned for such an application. The 180-grain bullet is the most useful for most big-game hunting situations.

The 220-grain bullet can be used on the biggest North American game including brown and grizzly bear.

It remains an excellent choice for the serious big-game hunter who can handle the recoil.

.300 Winchester Magnum
150-Grain Bullet—Light Big Game, Big Game
180-Grain Bullet—Big Game
220-Grain Bullet—Big Game

A great many hunters use the 150-grain bullet for deer-sized animals. Of course, they are overgunned in so doing. However, for the one-gun big-game hunter who wants to be able to hunt all North American big game, the .300 Winchester Magnum is an excellent choice.

The 180-grain bullet will work well on all but our biggest bears. For the big bears, the 220-grain bullet gets the nod. Accuracy, with capable shooters, is excellent. This is not a cartridge for those who object to large helpings of recoil.

.30-06 Springfield
110-Grain Bullet—Varmint
125-Grain Bullet—Varmint
150-Grain Bullet—Light Big Game
180-Grain Bullet—Big Game
220-Grain Bullet—Big Game

Any good big-game cartridge will work well here.

The .30-06 is a perennial favorite with big-game hunters. It will handle game from deer, antelope, and sheep right on up to elk and moose. While more than a fair number of grizzly and brown bear have been taken with the 220-grain bullet, the .30-06 is just barely adequate for that job.

Accuracy is usually excellent. However, accuracy with the light varmint bullets suffers a bit, and 200 yards seem to be about it with such loads.

While this cartridge is no longer the standard U.S. service cartridge, its popularity has not slowed down.

Recoil is not severe for an experienced shooter, but it's getting up there; it's about as heavy as the "average shooter" can handle. For most, the .30-06 represents the heaviest practical cartridge for general big-game hunting.

For those who desire to obtain off-season varmint practice, the 125-grain bullet usually shoots a bit better than the 110-grain bullet.

.30-40 Krag
180-Grain Bullet—Light Big Game
220-Grain Bullet—N.R.

Originally one of the first repeating-rifle service cartridges of the U.S. military, the .30-40 Krag is now all but forgotten. Only one commercial model rifle has been chambered for it in recent years. With the 180-grain bullet, it makes a fine light big-game cartridge for medium ranges. The 220-grain bullet has no practical application. Its ballistics are not sufficient for bigger game, and the lighter bullet does a better job on deer-sized animals.

.300 Savage
150-Grain Bullet—Light Big Game
180-Grain Bullet—Light Big Game

The .300 is a superb cartridge for game up to 500 pounds. It offers ballistics that are quite superior to the .30-30 without going so far that recoil and noise become a problem. Skilled woodsmen have accounted for many a moose with the 180-grain bullet in a Savage Model 99 chambered for the .300 Savage.

This cartridge would be an excellent choice for the one-gun, light big-game hunter for any area where the ranges won't exceed 225 or so yards.

Accuracy is usually excellent in most .300 Savage combinations. Popularity is waning but not very rapidly.

.303 Savage
190-Grain Bullet—Light Big Game

This cartridge is a ballistic equal of the .30-30. The slight reduction in velocity is offset by the heavier bullet weight.

The cartridge never realized any real amount of popularity and is now among the dead cartridges.

.308 Winchester
110-Grain Bullet—Varmint
125-Grain Bullet—Varmint
150-Grain Bullet—Light Big Game
180-Grain Bullet—Big Game
200-Grain Bullet—Big Game

The .308 replaced the .30-06 as the official U.S. service cartridge. It has since been replaced by the diminutive 5.56mm as the official military round. As the military round, the .308 is known as the 7.62mm.

Accuracy is usually, gun for gun, superior to the .30-06. This is due mostly to the 1-12-inch twist used in the .308, as opposed to the 1-10-inch twist in the .30-06. The .308 is a much shorter case than the .30-06, but it produces nearly the same ballistics due to the new powders that were designed for it. Winchester Ball Powder made it possible for the .308 to turn in ballistics that average only approximately 100 fps slower than the .30-06.

The .308 has never gained quite as much popularity as the .30-06. Perhaps this was due to the far shorter time it has served as our official U.S. military round. It actually deserves, perhaps, more popularity than the .30-06. The 100 fps difference is never noticeable in the field. The additional accuracy and the shorter-length actions required give it several advantages over the .30-06. Like the .30-06, the .308 is at home from deer to everything but our biggest bear.

Due to its increased accuracy, it can be used on varmints to about 225 yards. Deservedly, the .308 will continue to be a highly favored cartridge. The .308 may well be the best choice for the big-game hunter who is interested in varmints only to keep in shape.

.31 CALIBER

.303 British
150-Grain Bullet—Light Big Game
180-Grain Bullet—Light Big Game

The .303 British was the long-time military round of Great Britain. This cartridge has little or no features that would enable it to claim fame. Most of the European arms for this caliber have badly worn barrels due to the cordite powder and corrosive priming used by the British for loading this round. When a .303 British rifle can be found with a good bore, it will serve as a most adequate light big-game cartridge.

.32 CALIBER (8mm)

8mm Mauser (8×57mm)
170-Grain Bullet—Light Big Game
A great number of surplus military rifles chambered for the 8mm entered the country prior to 1968. These guns, when properly sporterized, make adequate light big-game rifles if the barrel is not eroded or corroded. A few commercial sporters made in the U.S. were chambered for the 8mm Mauser.

.32 Remington
170-Grain Bullet—Light Big Game
The .32 Remington is a rimless version of the .32 Winchester Special. Ballistics are almost identical to the .30-30, and hence, it is a suitable cartridge for deer to approximately 200 yards.

The .32 Remington is all but forgotten by today's shooters.

.32 Winchester Special
170-Grain Bullet—Light Big Game
The .32 Winchester Special is the ballistic twin of the .30-30 but not quite as accurate. No firearms are currently chambered for this cartridge, and it is on the list of dying cartridges.

The only purpose the .32 Winchester Special ever served has not been valid for many years. When the .30-30 was first introduced as a smokeless cartridge, reloaders were only able to obtain black powder. Due to the twist rate of the .30-30 and the lower velocities of black-powder loads, reloading produced poor accuracy. The .32 Winchester was introduced with a twist rate that would provide acceptable accuracy with black powder.

.32-20 Winchester
100-Grain Bullet—N.R.
Because it is too big for small game and too small for big game, the .32-20 is left in a useless limbo. With the now discontinued 80-grain, high-speed loading, it was useful on chuck to about 100 yards. The current 100-grain lead-bullet load can be used for short ranges (50 yards or less) on varmint, but head shots are a must.

.32-40 Winchester
170-Grain Bullet—N.R.
The .32-40 is dead, because it was too slow to be a useful deer cartridge and too big for small game. It was a fine accurate cartridge in its day, and it was favored by many target shooters. Most lead-bullet shooters still have a warm spot in their hearts for this old has-been.

.33 CALIBER

.338 Winchester Magnum
200-Grain Bullet—Big Game
250-Grain Bullet—Big Game
300-Grain Bullet—Dangerous Game

The .338 is a powerful cartridge that can be used on any big-game animal. The 300-grain load can be used on some dangerous game, but it is too light for elephant, rhino, and Cape buffalo.

Accuracy is good, but recoil is quite heavy.

Anyone who hunts large big-game animals frequently and can handle heavy recoil will never find himself lacking for power when using this cartridge.

.34 CALIBER

.348 Winchester
200-Grain Bullet—Big Game

The .348 is an excellent medium-range (150 yards or less), big-game cartridge. Only one firearm—the Model 71 Winchester—was ever chambered for this cartridge, and it has long been discontinued. The obsolescence of the 150-grain and 250-grain bullets was due to a lack of popularity. The 71's, today, are worth more as collector's items than as working firearms, and applications of the cartridge, therefore, are academic.

.35 CALIBER

.35 Remington
150-Grain Bullet—Light Big Game
200-Grain Bullet—Light Big Game

The .35 Remington makes an excellent medium-range deer cartridge. Its ballistics are quite similar to the .30-30. Some feel quite strongly that it performs somewhat better than the .30-30, but that is hogwash. Either will do the job nicely.

The 150-grain loading by Remington is a spitzer-shaped bullet and hence, it is the author's opinion that it should not be used in tubular magazines with perhaps the exception of the spiraled Remington tubes. In any event the 200-grain bullet will outperform the 150-grain bullet.

.350 Remington Magnum
200-Grain Bullet—Light Big Game, Big Game

This big cousin to the 6.5mm Remington Magnum also proved to be a loser from the very first with respect to popularity. However, it is quite a powerhouse in the 20-inch carbines made for it. Ballistics are adequate for any big game up to 200 yards, and for smaller animals the range can be stretched 300 or more yards.

Originally, a 250-grain bullet was also available but, due to poor popularity, the cartridge is now offered in only one bullet weight.

This is Winchester .458 Magnum country, although a highly skilled rifleman could get by with a .375 H&H Magnum on rhino.

The cartridge is unfortunately dying, perhaps because it was never chambered in a popular firearm.

.351 Winchester Self-Loader
  180-Grain Bullet—N.R.

The ballistics of this cartridge would make it marginal for small deer at 50 yards. It is dead, and it should, and will, stay so.

.358 Winchester
  200-Grain Bullet—Light Big Game
  250-Grain Bullet—Big Game

Although this powerhouse deserved a better fate than it received, it never really caught on. At this point, it is a rapidly dying cartridge. Savage currently chambers the only commercial gun available for this cartridge in the Model 99.

In bolt-action rifles, accuracy was always good. It will handle all North American big game with the exception of the larger bears.

.37 CALIBER

.375 H&H Magnum
270-Grain Bullet—Big Game, Dangerous Game
300-Grain Bullet—Big Game, Dangerous Game

If there ever was one cartridge that a hunter could use the world over for all big-game situations, the .375 H&H Magnum is it.

It is an extremely accurate cartridge, but it is not for everyone. The recoil is nothing less than brutal. For those who can handle it, the .375 will take everything from deer to elephant. Yes, it's more gun than needed on animals of less than 1000 pounds, but it is the very best all-around, big game/dangerous game rifle that has ever come down the pike. The 300-grain full metal case is the proper choice for elephant and Cape buffalo. The 270-grain soft point will do the entire remaining job.

The .375 is truly one of the all-time cartridge greats. Its ballistics make it an excellent long-range big-game rifle. The .375 H&H Magnum will be with us for a long, long time. It is possible that it will never be exceeded as the ideal choice for the one-gun, worldwide big-game hunter.

.38-55 Winchester
255-Grain Bullet—N.R.

The .38-55 is just barely adequate for short-range deer hunting. The cartridge is dead, and no reason exists to change this posture.

.40 CALIBER

.38-40 Winchester
180-Grain Bullet—N.R.

The .38-40 is hardly adequate for short-range deer hunting, and it hasn't any other practical application. The cartridge is justifiably forgotten.

.44-40 Winchester
200-Grain Bullet—N.R.

The .44-40 has killed a lot of deer in its day, but its ballistics are marginal for anything except the shortest ranges. There are a great number of cartridges which should be chosen in preference to this reminder of yesteryear.

.44 Remington Magnum
240-Grain Bullet—Light Big Game

It is with reservation that this handgun cartridge is listed for light big game. On small deer, at ranges not exceeding 100 yards, it will get the job done, but only if care is taken to use well-placed shots.

.444 Marlin
240-Grain Bullet—Light Big Game

This is an excellent cartridge for light big game to 200 yards. While not very popular, it nonetheless is a quite satisfactory performer. The blunt-nose bullet has quite a bit of energy at the shorter ranges, but due to its profile, ballistics fall off very rapidly.

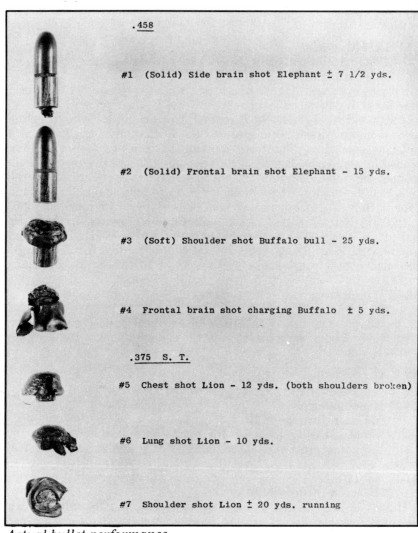

.458

#1 (Solid) Side brain shot Elephant ± 7 1/2 yds.

#2 (Solid) Frontal brain shot Elephant - 15 yds.

#3 (Soft) Shoulder shot Buffalo bull - 25 yds.

#4 Frontal brain shot charging Buffalo ± 5 yds.

.375 S. T.

#5 Chest shot Lion - 12 yds. (both shoulders broken)

#6 Lung shot Lion - 10 yds.

#7 Shoulder shot Lion ± 20 yds. running

Actual bullet performance.

.45 CALIBER

.45-70 Government
405-Grain Bullet—Light Big Game
The .45-70 is suitable to about 125 yards for light big game. The poor trajectory of the cartridge makes shooting past 125 yards highly impractical. However, as a short-range, big-game gun it does an excellent job.

For hostile Cape buffalo, the .458 is the best choice.

.458 Winchester Magnum
    500-Grain Bullet—Dangerous Game
    510-Grain Bullet—Dangerous Game

The .458 is actually too heavy a cartridge to have any application on American big game. For the stout of heart, it could be used on large brown bear. Basically it is a cartridge for elephant, rhino, and buffalo. Recoil is extremely heavy, and few can master the cartridge.

While elephants have been taken with lesser cartridges, the 458 Winchester Magnum is ideal for such situations.

CHAPTER FIVE

Centerfire Exterior Ballistics Tables for Rifles

Once again the reader is reminded that, as with the rimfire ballistics tables, any centerfire ballistics table is based on an averaging of all factors concerned and the extension of these factors by computer to develop the total tables.

Variations from firearm to firearm and from one lot of ammunition to the next can cause results which would deviate from the tables. In centerfire high-velocity cartridges, velocity variations from extreme high to extreme low in a given ten-shot string could easily be as much as 75 fps. Such a change would obviously affect all the downrange ballistics to some extent. Therefore, the numbers in the tables should not be considered as absolutes. They represent the average results you can anticipate.

Ballistics tables, at best, are representative tables determined by an *average* velocity, an *average* bullet weight, an *average* bore diameter, an *average* powder charge, etc.

Exterior ballistics of centerfire rifle cartridges are affected by the barrel length of the firearm. The extensive exterior ballistics tables in this chapter have been developed with a standard barrel length of 24 inches. The following cartridges are the exceptions:

Caliber	Barrel Length Used to Develop Ballistics
.30 Carbine	18″
.350 Remington Magnum	20″
.44 Remington Magnum	20″

To estimate the differences in muzzle velocity due to barrel-length changes, the following table will prove to be a useful guideline. This chart would apply to a range of barrel lengths from 20 inches to 26 inches.

The heart of a ballistics testing lab is the universal receiver used to determine pressure measurements.

Muzzle Velocity Range fps	Approximate Change in Velocity for Each 1" Change in Barrel Length from the Nominal
up to 2000	5 fps
2001 to 2500	10 fps
2501 to 3000	20 fps
3001 to 3500	30 fps
3501 to 4000	40 fps

Due to the great amount of data contained in the exterior ballistics tables for rifles, it was impossible to list all the data for a given cartridge on one page. Therefore, the use of the tables involves a line code for each group of pages.

To use the tables, simply locate the cartridge and load of your choice. Note that a line number is listed to the right of each cartridge and load. Simply refer to that line number on the following tables to locate the ballistics that apply to your cartridge.

The ballistics tables include data for "hold under" for a 45-degree hill. This data may be computed to any other angle by using the following formula:

$$h \left(\frac{N^2}{45} \right) = \text{Hold under for new angle}$$

In which:

h = Hold under for a 45-degree angle. Obtain this number from ballistic tables for the caliber and bullet being used and under the column for the range desired.

N = New angle for which you wish to calculate hold under.

For example:

What is the hold under for the 6mm Remington using an 80-grain bullet at 3470 fps for a 30-degree angle at 100 yards? Referring to line 1 of our exterior ballistics tables we find that the hold under for this combination (line 1 in tables) is .5 inches for a 45-degree angle at 100 yards. Therefore:

$$h \left(\frac{N^2}{45} \right) = \text{Hold under}$$

$$.5 \left(\frac{30^2}{45} \right) = \text{Hold under}$$

$.5 (.6666)^2 = \text{Hold under}$
$.5 \times .4444 = \text{Hold under}$
$.2'' = \text{Hold under for new angle of 30 degrees}$

Please remember that the hold under distance you arrive at from either the ballistics tables or your calculations does not mean that you would aim that distance below your target. Rather it means that you would subtract the distance arrived at from the required normal hold over.

Let's take a hypothetical case. Your rifle is sighted-in at 100 yards and hits 5 inches low at 200 yards. You now desire to hit a target at a 200-yard distance. On flat land you hold over 5 inches. However, in this hypothetical case you are 200 yards from

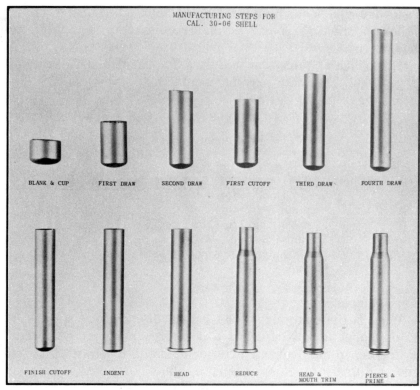

MANUFACTURING STEPS FOR
CAL. 30-06 SHELL

BLANK & CUP   FIRST DRAW   SECOND DRAW   FIRST CUTOFF   THIRD DRAW   FOURTH DRAW

FINISH CUTOFF   INDENT   HEAD   REDUCE   HEAD & MOUTH TRIM   PIERCE & PRIME

Steps in the manufacture of a .30-06 shell.

your target, but the target is up- or downhill from you at a 45-degree angle. Assume the hold under from the tables is 3 inches. The 3-inch hold under is subtracted from the required flat land hold over of 5 inches, thus your actual required hold over is 2 inches.

Additionally, you can calculate the height of trajectory above the line of sight whenever your sight height is not equal to the sight height used in the ballistics tables (.9 inches above the centerline of the bore). To do so, use the following formula:

$$\left[\left(\frac{R_t}{R_s}\right) - 1\right](h - 0.9) = \text{New height of trajectory to be added to original height of trajectory}$$

Where:

R_t = Target range in yards
R_s = Sight-in range in yards
h   = New height of sights above bore line in inches

For example:

What would the new height of trajectory be for a 6mm Remington using an 80-grain bullet at 3470 fps at 100 yards, when sighted-in at 150 yards with sights 2 inches above bore line?

Referring to the ballistics tables we find on line 1 under "height of trajectory (inches) above line of sight (do not use "midrange trajectory height in inches from bore line") with sights .09 inch above bore line" that the original height of trajectory was .6 inch at 100 yards with zero at 150 yards. Then to find the new trajectory at 100 yards with sights 2.0 inches above bore line we have:

$$\left[\left(\frac{R_t}{R_s}\right) - 1\right](h - 0.9) = \text{New height of trajectory to be added to the original .6"}$$

$$\left[\left(\frac{100}{150}\right) - 1\right](2.0 - 0.9) = \text{New height of trajectory to be added to the original .6"}$$

$(-.6666) - 1 (1.1) = $ New height of trajectory to be added to the original .6"

$(-.3333) (1.1) = $ New height of trajectory to be added to the original .6"

$-.36 = $ New height of trajectory to be added to the original .6"

$-.36 + .6 = .23"$ New height of trajectory for sight 2" above bore instead of .09" above bore line as in tables

The conditions that apply to the ballistics tables are:

Temperature = 59.0 degrees F.
Barometer = 29.53 inches of mercury
Elevation = 0.0 feet

Maximum-range tables are estimates based on normal conditions disregarding the effects of wind and variations in velocity as they occur from shot to shot. Additionally, these estimates are based on a flat-based, pointed bullet of .8-caliber ogive. Under certain circumstances it is possible that the range shown could be exceeded. If such an occurrence would create a risk of injury, an appropriate allowance should be made.

CARTRIDGE TYPES

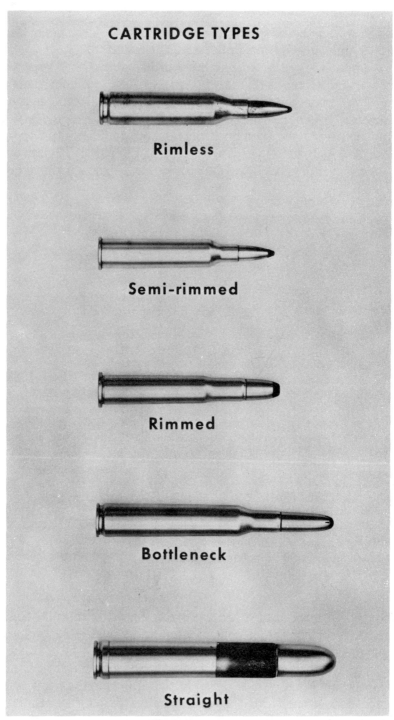

Rimless

Semi-rimmed

Rimmed

Bottleneck

Straight

Cartridge types.

Important Note:

The ammunition industry has just recently adjusted the nominal muzzle velocity on a number of cartridges.

The ballistics tables contained in the book are the first listings of these velocity adjustments. Therefore, these tables will appear to be different in some areas than any previous tables you may have read.

The cartridges affected are as follows:

Cartridge	Bullet Weight in Grains	Previous Velocity	New Velocity*
6mm Remington	90	3240	3175
7mm Mauser	175	2450	2420
8mm Mausr	170	2490	2340
.243 Winchester	75	3400	3325
.243 Winchester	80	3400	3325
.25-06 Remington	120	3040	3000
.25-35 Winchester	117	2255	2210
.264 Win. Mag.	100	3595	3300
.264 Win. Mag.	140	3125	3015
.300 Win. Mag.	180	2990	2950
.303 British	180	2505	2450
.308 Winchester	110	3250	3150
.308 Winchester	125	3080	3030
.32-20 Winchester	100	1275	1200
.35 Remington	150	2365	2275
.35 Remington	200	2055	2000
.38-40 Winchester	180	1315	1150
.44-40 Winchester	200	1295	1175
.458 Win. Mag.	500	2105	2025
.458 Win. Mag.	510	2100	2025

*Velocities shown here are instrumental at 15 feet. They are corrected to velocity at the muzzle in the ballistics tables.

EXTERIOR BALLISTICS TABLES—RIFLE

Line No.	Cartridge	Wgt. Grs.	Ballistic Coefficient	Bullet Type				Bullet Drop-Inches from Bore Line				
				I.V.I.	Federal	Rem-Peters	W-W	100	200	300 Yards	400	500
1	6mm Remington	80	0.255			PSP,HPL	PSP	1.57	6.86	16.98	33.43	58.33
2	6mm Remington	90	0.299			PSPCL		1.84	7.93	19.37	37.57	64.46
3	6mm Remington	100	0.356			PSPCL	PP(SP)	1.89	8.04	19.36	36.96	62.27
4	6.5mm M. Sch.	160	0.359	SP				4.63	19.99	48.80	94.46	161.08
5	6.5×55mm	160	0.359	SP				3.25	13.99	33.96	65.45	111.37
6	6.5mm Rem. Mag.	100	0.260			PSPCL		1.64	7.17	17.73	34.86	60.74
7	6.5mm Rem. Mag.	120	0.324			PSPCL		1.80	7.74	18.76	36.08	61.31
8	7mm Mauser	139	0.311	PSP				2.65	11.46	28.07	54.67	94.17
9	7mm Mauser	139	0.330		HSSP			2.63	11.36	27.66	53.54	91.58
10	7mm Mauser	160	0.274	SP				2.99	13.12	32.67	64.81	113.84
11	7mm Mauser	175	0.273	KK	HSSP	SP	SP	3.19	14.04	35.05	69.68	122.60
12	7mm Rem. Mag.	125	0.292			PSPCL		1.71	7.38	18.05	35.04	60.17
13	7mm Rem. Mag.	150	0.346	ST	HSSP	PSPCL	PP(SP)	1.91	8.18	19.74	37.79	63.86
14	7mm Rem. Mag.	175	0.427			PSPCL		2.24	9.46	22.53	42.52	70.76
15	7mm Rem. Mag.	175	0.273	SP	HSSP		PP(SP)	2.31	10.09	24.98	49.24	86.05
16	8mm Mauser	170	0.205	PSP	HSSP	SPCL	PP(SP)	3.52	16.13	42.13	87.84	161.05
17	8mm Mauser*	170	0.312					2.98	12.92	31.71	61.89	106.82
18	.17 Remington	25	0.151			HPPL		1.23	5.73	15.31	33.08	64.66
19	.218 Bee	46	0.130			HP	OPE(HP)	2.55	12.60	36.34	84.38	168.46
20	.22 Hornet	45	0.130	PSP		SP,HP	SP	2.89	14.42	41.80	96.39	189.64
21	.22 Hornet	46	0.130				OPE(HP)	2.89	14.41	41.77	96.32	189.49
22	.22 Savage	70	0.249	PSP				2.50	11.07	27.77	55.60	93.84
23	.222 Remington	50	0.188			HPPL		1.98	9.04	23.51	49.18	92.00
24	.222 Remington	50	0.175	PSP	SP	PSP	PSP	2.00	9.21	24.28	51.60	98.18
25	.222 Rem. Mag.	55	0.209			HPPL		1.84	8.24	21.00	42.85	78.00
26	.22 Rem. Mag.	55	0.197			PSP		1.85	8.36	21.49	44.33	21.74
27	.223 Remington	55	0.209			HPPL	PSP	1.84	8.24	21.00	42.85	78.00
28	.223 Remington	55	0.197		SP	PSP	PSP	1.85	8.36	21.50	44.37	61.82
29	.225 Winchester	55	0.208					1.51	6.75	17.13	34.72	63.75
30	.22-250 Remington	55	0.230			HPPL		1.37	6.05	15.11	30.09	53.19
31	.22-250 Remington	55	0.197	PSP	SP	PSP		1.39	6.25	15.96	32.58	59.36
32	.243 Winchester	75	0.236	PSP		PSP		1.70	7.49	18.73	37.36	66.21
33	.243 Winchester	80	0.255		SP	PSP		1.69	7.37	18.27	36.02	63.00

Line No.	Cartridge	Grains	B.C.	Bullet				100	200	300	400	500
34	.243 Winchester	80	0.255	PSP				1.62	7.07	17.50	34.47	60.21
35	.243 Winchester	100	0.356		HSSP		PP(SP)	2.11	9.01	21.73	41.57	70.19
36	.25-06 Remington	87	0.263	HPPL PSPCL		HPPL		1.59	6.95	17.13	33.62	58.43
37	.25-06 Remington	87	0.230					1.61	7.13	17.87	35.72	63.47
38	.25-06 Remington	90	0.259		HP		PSP	1.60	6.97	17.21	33.82	58.91
39	.25-06 Remington	100	0.292					1.79	7.76	18.98	36.88	63.41
40	.25-06 Remington	117	0.349		HSSP		PSPCL	1.98	8.45	20.37	38.99	65.68

*Discontinued

EXTERIOR BALLISTICS TABLES—RIFLE

Line No.	Cartridge	Velocity in fps at Yards						Energy in ft.-lbs. at Yards						Time of Flight in Seconds Yards				
		0	100	200	300	400	500	0	100	200	300	400	500	100	200	300	400	500
1	6mm Remington	3470	3064	2694	2352	2036	1747	2139	1667	1289	982	736	542	0.092	0.156	0.316	0.453	0.612
2	6mm Remington	3190	2886	2558	2273	2007	1760	2033	1638	1307	1032	805	619	0.099	0.210	0.334	0.475	0.635
3	6mm Remington	3130	2857	2600	2357	2127	1911	2175	1812	1501	1233	1004	811	0.100	0.210	0.332	0.466	0.614
4	6.5mm M. Sch.	2010	1803	1611	1439	1289	1167	1435	1155	922	736	590	484	0.158	0.334	0.531	0.751	0.996
5	6.5×55mm	2390	2160	1944	1741	1555	1389	2029	1657	1342	1077	859	685	0.132	0.278	0.442	0.624	0.828
6	6.5mm Rem. Mag.	3390	2998	2639	2308	2000	1720	2551	1995	1546	1183	888	657	0.094	0.201	0.322	0.462	0.624
7	6.5mm Rem. Mag.	3210	2905	2621	2353	2102	1867	2745	2248	1830	1475	1177	929	0.098	0.207	0.328	0.463	0.614
8	7mm Mauser	2660	2380	2117	1871	1646	1444	2183	1748	1383	1080	836	643	0.119	0.253	0.404	0.575	0.769
9	7mm Mauser	2680	2395	2145	1911	1694	1498	2183	1770	1420	1127	886	692	0.119	0.251	0.399	0.566	0.755
10	7mm Mauser	2520	2213	1928	1660	1438	1248	2256	1740	1320	988	735	553	0.127	0.272	0.448	0.633	0.858
11	7mm Mauser	2440	2137	1857	1603	1382	1204	2313	1774	1340	998	742	563	0.131	0.282	0.456	0.658	0.891
12	7mm Rem. Mag.	3310	2966	2647	2350	2073	1815	3040	2441	1944	1533	1193	914	0.096	0.203	0.323	0.459	0.614
13	7mm Rem. Mag.	3110	2830	2568	2320	2085	1866	3221	2667	2196	1792	1448	1160	0.101	0.212	0.335	0.472	0.624
14	7mm Rem. Mag.	2860	2645	2440	2244	2057	1879	3178	2718	2313	1956	1644	1372	0.109	0.227	0.355	0.495	0.648
15	7mm Rem. Mag.	2860	2528	2219	1933	1671	1440	3178	2483	1913	1452	1085	806	0.112	0.238	0.383	0.550	0.744
16	8mm Mauser	2360	1969	1622	1333	1123	997	2102	1463	993	671	476	375	0.139	0.307	0.512	0.758	1.043
17	8mm Mauser*	2510	2240	1987	1752	1539	1353	2378	1894	1490	1158	894	691	0.127	0.269	0.430	0.612	0.820
18	.17 Remington	4040	3284	2644	2086	1606	1235	906	599	388	242	143	85	0.082	0.184	0.312	0.476	0.690
19	.218 Bee	2860	2188	1619	1198	980	863	835	489	268	147	98	76	0.120	0.280	0.497	0.777	1.104
20	.22 Hornet	2690	2042	1502	1128	947	840	723	417	225	127	90	70	0.128	0.300	0.533	0.826	1.163
21	.22 Hornet	2690	2042	1502	1128	948	840	739	426	230	130	92	72	0.128	0.300	0.532	0.825	1.162
22	.22 Savage	2760	2404	2076	1775	1509	1286	1184	898	670	490	354	257	0.116	0.251	0.407	0.591	0.806
23	.222 Remington	3140	2635	2182	1777	1432	1172	1094	771	529	351	228	152	0.104	0.225	0.382	0.570	0.803
24	.222 Remington	3140	2602	2123	1700	1350	1107	1094	752	500	321	202	136	0.105	0.233	0.391	0.589	0.836
25	.222 Rem. Mag.	3240	2773	2352	1969	1627	1341	1282	939	675	473	323	220	0.100	0.218	0.357	0.525	0.728

EXTERIOR BALLISTICS TABLES—RIFLE (continued)

Line No.	Cartridge	Velocity in fps at (Yards)						Energy in ft.-lbs. at (Yards)						Time of Flight in Seconds (Yards)				
		0	100	200	300	400	500	0	100	200	300	400	500	100	200	300	400	500
26	.222 Rem. Mag.	3240	2748	2305	1906	1556	1272	1282	922	649	444	296	198	0.101	0.220	0.363	0.537	0.751
27	.223 Remington	3240	2773	2352	1969	1627	1341	1282	939	675	473	323	200	0.100	0.218	0.357	0.523	0.728
28	.223 Remington	3240	2747	2304	1905	1554	1270	1282	921	645	443	295	197	0.101	0.220	0.363	0.538	0.752
29	.225 Winchester	3570	3066	2616	2208	1838	1514	1556	1148	836	595	412	280	0.091	0.197	0.322	0.471	0.651
30	.22-50 Remington	3730	3253	2826	2436	2079	1755	1699	1292	975	725	520	376	0.086	0.185	0.300	0.433	0.590
31	.22-250 Remington	3730	3180	2695	2257	1863	1519	1699	1235	887	622	424	282	0.087	0.190	0.311	0.458	0.636
32	.243 Winchester	3350	2924	2536	2180	1854	1563	1869	1424	1071	791	572	407	0.096	0.206	0.334	0.483	0.659
33	.243 Winchester	3350	2955	2593	2259	1951	1670	1993	1551	1194	906	676	495	0.095	0.204	0.327	0.471	0.637
34	.243 Winchester	3420	3019	2652	2313	2000	1715	2077	1619	1279	950	710	522	0.093	0.199	0.321	0.460	0.622
35	.243 Winchester	2960	2697	2449	2215	1993	1786	1945	1615	1332	1089	882	708	0.106	0.223	0.352	0.495	0.654
36	.25-06 Remington	3440	3048	2690	2358	2051	1769	2286	1794	1398	1074	812	604	0.093	0.197	0.317	0.453	0.611
37	.25-06 Remington	3440	2995	2591	2222	1884	1583	2286	1733	1297	954	686	484	0.094	0.200	0.326	0.473	0.647
38	.25-06 Remington	3440	3043	2680	2344	2034	1749	2364	1850	1435	1098	827	611	0.093	0.198	0.318	0.455	0.614
39	.25-06 Remington	3230	2893	2580	2287	2014	1762	2316	1858	1478	1161	901	689	0.098	0.208	0.332	0.471	0.631
40	.25-06 Remington	3060	2786	2528	2284	2054	1838	2432	2016	1660	1355	1096	877	0.103	0.216	0.341	0.479	0.638

*Discontinued

EXTERIOR BALLISTICS TABLES—RIFLE

Line No.	Cartridge	Midrange Trajectory Height in Inches from Bore Line (Yards)					Drift in Inches in a 10 mph Wind (Yards)					45° Angle Up- or Downhill Hold (Inches) under Target (Yards)				
		100	200	300	400	500	100	200	300	400	500	100	200	300	400	500
1	6mm Remington	0.4	1.9	4.8	9.9	17.9	1.0	4.1	9.9	18.8	31.6	0.5	2.0	5.0	9.8	17.1
2	6mm Remington	0.5	2.1	5.4	10.9	19.3	0.9	3.9	9.2	17.4	29.0	0.5	2.3	5.7	11.0	18.9
3	6mm Remington	0.5	2.1	5.3	10.4	18.1	0.8	3.3	7.8	14.5	23.8	0.6	2.4	5.7	10.8	18.2
4	6.5mm M. Sch.	1.2	5.4	13.6	27.2	48.0	1.5	6.2	14.6	27.2	44.0	1.4	5.9	14.3	27.7	47.2
5	6.5×55mm	0.8	3.7	9.4	18.7	33.0	1.1	4.8	11.4	21.4	35.3	1.0	4.1	9.9	19.2	32.6
6	6.5mm Rem. Mag.	0.4	1.9	5.0	10.3	18.6	1.0	4.2	10.0	19.0	31.9	0.5	2.1	5.2	10.2	17.8
7	6.5mm Rem. Mag.	0.5	2.1	5.2	10.3	18.1	0.8	3.5	8.3	15.6	25.8	0.5	2.3	5.5	10.6	18.0

No.	Cartridge															
8	7mm Mauser	0.7	3.1	7.8	15.9	28.4	1.1	4.8	11.5	21.7	36.2	0.8	3.4	8.2	16.0	27.6
9	7mm Mauser	0.7	3.0	7.7	15.4	27.3	1.1	4.5	10.7	20.2	33.6	0.8	3.3	8.1	15.7	26.8
10	7mm Mauser	0.8	3.6	9.3	19.3	35.4	1.4	6.0	14.5	27.7	46.2	0.9	3.8	9.6	19.0	33.3
11	7mm Mauser	0.8	3.8	10.0	20.8	38.2	1.5	6.3	15.3	29.2	48.6	0.9	4.1	10.3	20.4	35.9
12	7mm Rem. Mag.	0.4	2.0	5.0	10.1	18.1	0.9	3.8	9.0	17.0	28.3	0.5	2.2	5.3	10.0	17.6
13	7mm Rem. Mag.	0.5	2.2	5.4	10.7	18.7	0.8	3.4	8.1	15.1	24.9	0.6	2.4	5.8	11.1	18.7
14	7mm Rem. Mag.	0.6	2.5	6.1	11.8	20.2	0.7	3.1	7.2	13.3	21.7	0.7	2.8	6.6	12.5	20.7
15	7mm Rem. Mag.	0.6	2.7	7.1	14.5	26.5	1.2	5.0	12.0	23.0	38.6	0.7	3.0	7.3	14.4	25.2
16	8mm Mauser	0.9	4.5	12.6	27.8	53.4	2.1	9.3	22.9	43.9	71.7	1.0	4.7	12.3	25.7	47.2
17	8mm Mauser*	0.8	3.5	8.9	18.0	32.3	1.2	5.2	12.5	23.6	39.2	0.9	3.8	9.3	18.1	31.3
18	.17 Remington	0.3	1.6	4.7	10.8	22.6	1.4	6.3	15.7	31.5	56.2	0.4	1.7	4.5	9.7	18.9
19	.218 Bee	0.7	3.8	11.8	29.6	62.0	2.7	12.3	32.0	62.9	102.0	0.7	3.7	10.6	24.7	49.3
20	.22 Hornet	0.8	4.3	13.7	33.8	69.3	2.9	13.5	34.9	66.8	106.5	0.8	4.2	12.2	28.2	55.5
21	.22 Hornet	0.8	4.3	13.7	33.7	69.3	2.9	13.5	34.8	66.7	106.4	0.8	4.2	12.2	28.2	55.5
22	.22 Savage	0.7	3.0	8.0	16.7	31.2	1.4	5.9	14.3	27.4	46.3	0.7	3.2	8.1	16.3	28.9
23	.222 Remington	0.5	2.5	7.0	15.5	30.8	1.5	6.8	16.8	33.1	57.3	0.6	2.6	6.9	14.4	26.9
24	.222 Remington	0.5	2.6	7.3	16.6	33.5	1.7	7.8	18.3	36.4	63.1	0.6	2.7	7.1	15.1	28.8
25	.222 Rem. Mag.	0.5	2.3	6.1	13.2	25.3	1.3	5.7	14.0	27.2	46.7	0.5	2.4	6.2	12.6	22.8
26	.222 Rem. Mag.	0.5	2.3	6.3	13.8	26.9	1.4	6.1	15.0	29.4	50.7	0.5	2.4	6.3	13.0	23.9
27	.223 Remington	0.5	2.3	6.1	13.2	25.3	1.3	5.7	14.0	27.2	45.7	0.5	2.4	6.2	12.6	22.8
28	.225 Remington	0.5	2.3	6.3	13.8	27.0	1.4	6.1	15.0	29.4	50.8	0.5	2.4	6.3	13.0	24.0
29	.226 Winchester	0.4	1.9	5.0	10.6	20.2	1.2	5.0	12.2	23.7	40.5	0.4	2.0	5.0	10.2	18.4
30	.22-250 Remington	0.4	1.7	4.3	9.0	16.6	1.0	4.3	10.2	19.6	33.0	0.4	1.8	4.4	8.8	15.6
31	.22-250 Remington	0.4	1.7	4.7	10.0	19.3	1.2	5.1	12.3	23.9	41.2	0.4	1.8	4.7	9.5	17.4
32	.243 Winchester	0.4	2.0	5.4	11.2	20.8	1.1	4.7	11.4	22.0	37.2	0.5	2.2	5.5	10.9	19.4
33	.243 Winchester	0.4	2.0	5.2	10.6	19.4	1.0	4.3	10.4	19.8	33.3	0.5	2.2	5.4	10.6	18.5
34	.243 Winchester	0.4	1.9	4.9	10.2	18.5	1.0	4.2	10.1	19.2	32.3	0.5	2.1	5.1	10.1	17.6
35	.243 Winchester	0.5	2.4	6.0	11.8	20.5	0.9	3.6	8.4	15.7	25.8	0.6	2.6	6.4	12.2	20.6
36	.25-06 Remington	0.4	1.9	4.8	9.9	17.9	1.0	4.1	9.7	18.3	30.7	0.5	2.0	5.0	9.8	17.1
37	.25-06 Remington	0.4	2.0	5.1	10.7	20.0	1.1	4.7	11.4	21.8	37.1	0.5	2.1	5.2	10.5	18.6
38	.25-06 Remington	0.4	1.9	4.9	9.9	18.1	1.0	4.1	9.8	18.7	31.3	0.5	2.0	5.0	9.9	17.3
39	.25-06 Remington	0.5	2.1	5.3	10.7	19.1	0.9	3.9	9.3	17.6	29.2	0.5	2.3	5.6	10.8	18.6
40	.25-06 Remington	0.5	2.2	5.6	11.0	19.3	0.8	3.5	8.2	15.3	25.2	0.6	2.5	6.0	11.4	19.3

*Discontinued

55

EXTERIOR BALLISTICS TABLES—RIFLE

Height of Trajectory (Inches) Above or Below Sight Line
for Sights 0.9 Inches Above Bore Line

Line No.	Cartridge	Short Range (Yards)						Long Range (Yards)						
		50	100	150	200	250	300	100	150	200	250	300	400	500
1	6mm Remington	0.3	0.6	0	-1.6	-4.5	-8.7	2.4	2.7	1.9	0	-3.3	-14.9	-35.0
2	6mm Remington	0.3	0.7	0	-1.9	-5.2	-9.9	1.7	1.4	0	-2.8	-7.0	-20.8	-43.3
3	6mm Remington	0.4	0.7	0	-1.9	-5.1	-9.7	1.7	1.4	0	-2.7	-6.8	-20.0	-40.8
4	6.5mm M. Sch.	0.7	0	-3.4	-9.8	-19.5	-33.1	2.3	0	-5.3	-13.9	-26.3	-64.1	-122.9
5	6.5×55m	0.4	0	-2.3	-6.6	-13.2	-22.4	1.5	0	-3.6	-9.4	-17.9	-43.7	-84.0
6	6.5mm Rem. Mag.	0.3	0.6	0	-1.7	-4.7	-9.1	2.5	2.8	2.0	0	-3.5	-15.5	-36.4
7	6.5mm Rem. Mag.	0.4	0.7	0	-1.8	-4.9	-9.5	2.7	3.0	2.1	0	-3.5	-15.5	-35.3
8	7mm Mauser	0.2	0	-1.8	-5.3	-10.7	-18.3	2.6	2.2	0	-4.1	-10.4	-30.6	-64.3
9	7mm Mauser	0.2	0	-1.7	-5.2	-10.5	-18.0	2.6	2.1	0	-4.0	-10.2	-29.9	-61.8
10	7mm Mauser	0.3	0	-2.1	-6.2	-12.7	-21.9	1.4		-3.4	-9.2	-17.7	-44.5	-88.3
11	7mm Mauser	0.4	0	-2.3	-6.8	-13.8	-23.7	1.5	0	-3.7	-10.0	-19.1	-48.1	-95.4
12	7mm Rem. Mag.	0.3	0.7	0	-1.8	-4.3	-9.2	2.6	2.9	2.0	0	-3.4	-15.3	-35.2
13	7mm Rem. Mag.	0.4	0.8	0	-1.9	-5.2	-9.9	1.7	1.5	0	-2.8	-7.0	-20.5	-42.1
14	7mm Rem. Mag.	0.6	0.9	-1.5	-2.3	-6.0	-11.3	2.0	1.7	0	-3.2	-7.9	-22.7	-45.8
15	7mm Rem. Mag.	0.2	0	0	-4.6	-9.4	-16.3	2.3	1.9	0	-3.7	-9.4	-28.2	-59.5
16	8mm Mauser	0.5	0	-2.7	-8.2	-17.0	-29.8	1.8	0	-4.5	-12.4	-24.3	-63.8	-130.7
17	8mm Mauser*	0.3	0	-2.1	-6.1	-12.3	-21.0	1.4	0	-3.3	-8.8	-16.9	-41.8	-81.5
18	.17 Remington	0.1	0.5	0	-1.5	-4.2	-8.5	2.1	2.5	1.9	0	-3.5	-17.0	-44.3
19	.218 Bee	0.2	0	-2.1	-6.6	-14.3	-26.9	1.4	0	-3.8	-11.0	-22.8	-66.0	-145.2
20	.22 Hornet	0.3	0	-2.4	-7.7	-16.8	-31.3	1.6	0	-4.5	-12.8	-26.4	-75.6	-163.4
21	.22 Hornet	0.3	0	-2.4	-7.7	-16.8	-31.3	1.6	0	-4.5	-12.8	-26.4	-75.5	-163.3
22	.22 Savage	0.2	0	-1.7	-5.2	-10.3	-18.5	2.6	2.2	0	-4.2	-10.7	-32.6	-69.8
23	.222 Remington	0.5	0.9	0	-2.4	-6.5	-13.1	2.1	1.8	0	-3.6	-9.5	-30.2	-68.1
24	.222 Remington	0.5	0.9	0	-2.5	-6.8	-13.7	2.2	1.9	0	-3.8	-10.0	-32.3	-73.8
25	.222 Rem. Mag.	0.4	0.8	0	-2.1	-5.3	-11.4	1.8	1.6	0	-3.2	-8.2	-25.5	-56.0
26	.222 Rem. Mag.	0.4	0.8	0	-2.2	-6.0	-11.8	1.9	1.6	0	-3.3	-8.5	-26.7	-59.5
27	.223 Remington	0.4	0.8	0	-2.1	-5.8	-11.4	1.8	1.6	0	-3.2	-8.2	-25.5	-56.0
28	.223 Remington	0.4	0.8	0	-2.2	-6.0	-11.8	1.9	1.6	0	-3.3	-8.5	-26.7	-59.6
29	.225 Wincheter	0.2	0.6	0	-1.7	-4.6	-9.0	2.4	2.8	2.0	0	-3.5	-16.3	-39.5
30	.22-250 Remington	0.2	0.5	0	-1.4	-4.0	-7.7	2.1	2.4	1.7	0	-3.0	-13.6	-32.4

Continuation table (rows 31–40; column headings appear on the preceding page):

Line No.	Cartridge			I.V.I.										
31	.22-250 Remington	0.5	0.2	PSP	0	-1.5	-4.3	-8.4	2.2	2.6	1.9	0	-3.3	-15.4	-37.7
32	.243 Winchester	0.7	0.3		0	-1.8	-5.0	-9.8	1.6	1.4	0	-2.7	-7.0	-21.5	-46.1
33	.243 Winchester	0.7	0.3	SP	0	-1.8	-4.9	-9.4	2.6	2.9	2.1	0	-3.6	-16.2	-37.9
34	.243 Wincheter	0.8	0.3		0	-1.7	-4.6	-9.0	2.5	2.8	2.0	0	-3.6	-15.4	-36.2
35	.243 Winchester	0.9	0.5	SP	0	-2.2	-5.8	-11.0	1.9	1.6	0	-3.1	-7.8	-22.6	-46.3
36	.25-06 Remington	0.6	0.3		0	-1.7	-4.5	-8.8	2.4	2.7	2.0	0	-3.3	-14.9	-34.8
37	.25-06 Remington	0.6	0.3	PSP	0	-1.7	-4.8	-9.3	2.5	2.9	2.1	0	-3.6	-16.4	-39.1
38	.25-06 Remington	0.6	0.3		0	-1.7	-4.5	-8.8	2.4	2.7	2.0	0	-3.4	-15.0	-35.2
39	.25-06 Remington	0.7	0.4		0	-1.9	-5.0	-9.7	1.6	1.4	0	-2.7	-6.9	-20.5	-42.7
40	.25-06 Remington	0.8	0.5		0	-2.0	-5.4	-10.3	1.8	1.5	0	-2.9	-7.3	-21.2	-43.4

*Discontinued

EXTERIOR BALLISTICS TABLES—RIFLE

Line No.	Cartridge	Wgt. Grs.	Ballistic Coefficient	I.V.I.	Bullet Type Federal	Rem-Peters	W-W	100	200	Bullet Drop-Inches from Bore Line 300 Yards	400	500
41	.25-06 Remington	120	0.362	PSP		PSPCL	PEP	2.04	8.69	20.91	39.92	67.25
42	.25-06 Remington	120	0.246				PP(SP)	2.04	9.01	22.50	44.83	79.34
43	.25-20 Winchester	86	0.190	SP		SP, Lead	SP, Lead	9.41	43.21	109.67	216.21	370.02
44	.25-35 Winchester	117	0.240			SPCL		3.88	17.44	44.49	90.36	166.65
45	.25-35 Winchester	117	0.213	SP			SP	3.94	17.99	46.76	96.80	175.64
46	.250 Savage	87	0.263				PSP	2.06	9.02	22.35	44.15	77.32
47	.250 Savage	100	0.285			PSP		2.37	10.31	25.41	49.86	86.64
48	.250 Savage	100	0.254	PSP			ST	2.39	10.54	26.34	52.52	92.92
49	.256 Win. Mag.	60	0.129				OPE(HP)	2.75	13.67	39.67	91.95	182.13
50	.257 Roberts	87	0.263				PSP	1.86	8.21	20.32	40.02	69.89
51	.257 Roberts	100	0.254				ST	2.26	9.94	24.79	49.33	87.11
52	.257 Roberts	117	0.240			SPCL	PP(SP)	2.73	12.14	30.68	61.91	110.90
53	.264 Win. Mag.	100	0.254			PSPCL	PSP	1.72	7.52	18.64	36.81	64.46
54	.264 Win. Mag.	140	0.385			PSPCL	PP(SP)	2.00	8.51	20.38	38.71	64.83
55	.270 Winchester	100	0.251			PSP	PSP	1.56	6.84	16.96	33.46	58.53
56	.270 Winchester	130	0.372		HSSP	BP	PP(SP)	1.91	8.10	19.44	36.99	62.09
57	.270 Winchester	130	0.307	PSP,ST		PSPCL	ST	1.92	8.21	19.86	38.10	64.57
58	.270 Winchester	150	0.345				PP(SP)	2.20	9.44	22.84	43.85	74.36
59	.270 Winchester	150	0.261		HSSP	SPCL		2.25	9.89	24.80	48.76	85.74

EXTERIOR BALLISTICS TABLES—RIFLE (continued)

Line No.	Cartridge	Wgt. Grs.	Ballistic Coefficient	I.V.I.	Federal	Rem-Peters	W-W	100	200	300 (Yards)	400	500
60	.270 Winchester	160	0.337	KKSP				2.53	10.89	26.47	51.07	87.08
61	.280 Remington	150	0.401			PSPCL		2.07	8.76	20.93	39.64	66.19
62	.280 Remington	165	0.290			SPCL		2.36	10.28	25.29	49.50	85.82
63	.284 Winchester	125	0.311				PP(SP)	1.89	8.14	19.82	38.30	65.45
64	.284 Winchester	150	0.344				PP(SP)	2.27	9.72	23.52	45.17	76.64
65	.30 Carbine	110	0.166		SP	SP	HSP,FMC	5.17	24.77	67.23	141.99	257.58
66	.30 Remington	170	0.254			SPCL	ST	4.28	19.13	48.47	97.61	172.89
67	.30-30 Winchester	150	0.218	SP,ST,PNEU	HSSP		OPE,PP,ST	3.41	15.45	39.90	82.30	149.78
68	.30-30 Winchester	150	0.193			SPCL	PP(SP),ST	3.46	15.99	42.22	89.07	164.82
69	.30-30 Winchester	170	0.254	KKSP,ST	HSSP	HPCL,SPCL	ST	3.97	17.71	44.81	90.26	160.21
70	.300 H&H Magnum	150	0.314				ST	1.90	8.19	19.91	38.44	65.61
71	.300 H&H Magnum	180	0.383			PSPCL	ST	2.22	9.44	22.66	43.11	72.37
72	.300 H&H Magnum	220	0.359				ST	2.79	11.94	28.90	55.53	94.21
73	.300 Win. Mag.	150	0.295		HSSP	PSPCL	PP(SP)	1.73	7.47	18.23	35.37	60.69
74	.300 Win. Mag.	180	0.436	ST	HSSP	PSPCL	PP(SP)	2.22	9.44	22.66	43.11	72.37
75	.300 Win. Mag.	220	0.380				ST	2.57	10.96	26.37	50.34	84.80
76	.300 Win. Mag.	220	0.294				PP(SP)	2.62	11.39	28.06	55.00	95.42
77	.30-06 Springfield	110	0.186				PSP	1.71	7.77	20.14	41.95	78.23
78	.30-06 Springfield	125	0.268		SP	PSP	PSP	1.91	8.35	20.63	40.57	70.72
79	.30-06 Springfield	150	0.365			BP		2.18	9.30	22.40	42.78	72.11
80	.30-06 Springfield	150	0.352			MC MIL		2.35	10.05	24.28	46.59	78.93

EXTERIOR BALLISTICS TABLES—RIFLE

Line No.	Cartridge	Velocity in fps at						Energy in ft.-lbs. at						Time of Flight in Seconds				
		0	100	200 (Yards)	300	400	500	0	100	200 (Yards)	300	400	500	100	200	300 (Yards)	400	500
41	.25-06 Remington	3010	2749	2502	2269	2048	1840	2414	2013	1668	1312	1117	902	0.104	0.219	0.345	0.482	0.638
42	.25-06 Remington	3050	2668	2316	1992	1698	1440	2478	1896	1429	1057	798	552	0.105	0.226	0.366	0.529	0.721

#	Cartridge																	
43	.25-20 Winchester	1.507	1.144	0.808	0.500	0.228	121	141	165	203	272	407	797	858	931	1030	1194	1460
44	.25-35 Winchester	1.031	0.758	0.511	0.317	0.146	281	355	483	676	943	1292	1041	1169	1363	1613	1905	2230
45	.25-35 Winchester	1.081	0.792	0.538	0.324	0.147	252	313	427	620	904	1292	984	1097	1282	1545	1866	2230
46	.250 Savage	0.705	0.522	0.363	0.225	0.105	437	595	801	1059	1380	1773	1504	1755	2036	2342	2673	3030
47	.250 Savage	0.743	0.552	0.386	0.240	0.113	474	630	832	1084	1392	1765	1461	1684	1936	2210	2504	2820
48	.250 Savage	0.780	0.572	0.396	0.244	0.114	398	547	751	1017	1351	1765	1339	1569	1839	2140	2467	2820
49	.256 Win. Mag.	1.143	0.809	0.519	0.292	0.125	95	122	176	317	586	1015	846	956	1148	1542	2097	2760
50	.257 Roberts	0.670	0.496	0.345	0.215	0.101	491	666	890	1171	1516	1941	1594	1857	2147	2462	2802	3170
51	.257 Roberts	0.754	0.554	0.383	0.237	0.111	427	588	805	1083	1433	1867	1387	1627	1904	2209	2541	2900
52	.257 Roberts	0.858	0.626	0.430	0.263	0.122	373	512	718	999	1363	1824	1199	1404	1663	1961	2291	2650
53	.264 Win. Mag.	0.645	0.476	0.331	0.206	0.096	600	821	1105	1461	1901	2447	1644	1923	2231	2565	2926	3320
54	.264 Win. Mag.	0.624	0.475	0.339	0.216	0.103	1139	1839	1682	2018	2406	2854	1914	2114	2326	2548	2782	3030
55	.270 Winchester	0.614	0.454	0.316	0.196	0.092	664	909	1219	1606	2088	2689	1730	2023	2343	2690	3067	3480
56	.270 Winchester	0.611	0.465	0.332	0.211	0.101	1087	1334	1622	1957	2343	2791	1941	2150	2371	2604	2849	3110
57	.270 Winchester	0.629	0.475	0.337	0.213	0.101	974	1226	1527	1883	2300	2791	1837	2061	2300	2554	2823	3110
58	.270 Winchester	0.675	0.509	0.361	0.228	0.109	973	1225	1528	1886	2307	2801	1709	1918	2142	2380	2632	2900
59	.270 Winchester	0.746	0.549	0.381	0.236	0.110	667	910	1235	1649	2165	2801	1415	1653	1926	2225	2550	2900
60	.270 Winchester	0.734	0.552	0.390	0.246	0.116	856	1088	1374	1719	2129	2609	1552	1750	1967	2200	2448	2710
61	.280 Remington	0.629	0.479	0.343	0.219	0.105	1223	1480	1777	2118	2509	2957	1916	2108	2310	2522	2745	2980
62	.280 Remington	0.738	0.549	0.384	0.240	0.113	801	1060	1391	1805	2308	2913	1479	1701	2949	2220	2510	2820
63	.284 Winchester	0.637	0.478	0.338	0.213	0.101	871	1121	1424	1788	2221	2736	1772	2010	2265	2538	2829	3140
64	.284 Winchester	0.686	0.517	0.367	0.232	0.110	940	1185	1480	1880	2243	2724	1680	1886	2108	2344	2595	2860
65	.30 Carbine	1.303	0.962	0.654	0.389	0.170	173	208	262	373	600	967	842	922	1035	1236	1567	1990
66	.30 Remington	1.059	0.783	0.540	0.331	0.153	405	592	666	913	1253	1696	1036	1153	1328	1555	1822	2120
67	.30-30 Winchester	1.003	0.730	0.495	0.299	0.137	357	461	651	904	1356	1902	1035	1177	1398	1684	2018	2390
68	.30-30 Winchester	1.060	0.768	0.515	0.307	0.138	315	399	565	858	1296	1902	973	1095	1303	1605	1973	2390
69	.30-30 Winchester	1.022	0.754	0.519	0.318	0.147	425	535	720	989	1355	1827	1061	1191	1381	1619	1875	2200
70	.300 H&H Magnum	0.638	0.479	0.338	0.213	0.101	1050	1347	1707	2188	2652	3262	1776	2011	2264	2534	2822	3130
71	.300 H&H Magnum	0.660	0.502	0.358	0.228	0.109	1292	1583	1927	2325	2785	3315	1798	1990	2196	2412	2640	2880
72	.300 H&H Magnum	0.760	0.573	0.407	0.257	0.122	1128	1415	1765	2183	2677	3251	1520	1702	1901	2114	2341	2580
73	.300 Win. Mag.	0.616	0.461	0.325	0.204	0.096	1095	1424	1827	2314	2900	3605	1813	2068	2342	2636	2951	3290
74	.300 Win. Mag.	0.660	0.502	0.358	0.228	0.109	1292	1584	1927	2325	2785	3315	1798	1991	2196	2412	2640	2880
75	.300 Win. Mag.	0.717	0.543	0.387	0.246	0.117	1314	1623	1993	2414	2927	3508	1640	1823	2020	2228	2448	2680
76	.300 Win. Mag.	0.779	0.579	0.405	0.253	0.119	961	1268	1664	2154	2774	3508	1403	1611	1846	2105	2383	2680
77	.30-06 Springfield	0.740	0.526	0.353	0.213	0.097	388	595	915	1356	1974	1790	1261	1561	1936	2365	2843	3380
78	.30-06 Springfield	0.673	0.499	0.348	0.217	0.102	706	953	1269	1662	2145	2736	1595	1853	2138	2447	2780	3140
79	.30-06 Springfield	0.661	0.501	0.357	0.226	0.108	1047	1295	1596	1944	2349	2820	1773	1974	2189	2416	2656	2910
80	.30-06 Springfield	0.695	0.525	0.372	0.236	0.112	923	1158	1442	1777	2170	2629	1665	1865	2081	2310	2553	2810

EXTERIOR BALLISTICS TABLES—RIFLE

Line No.	Cartridge	Midrange Trajectory Height in Inches from Bore Line					Drift in Inches in a 10 mph Wind					45° Angle Up- or Downhill Hold (Inches) under Target				
		100	200	300 Yards	400	500	100	200	300 Yards	400	500	100	200	300 Yards	400	500
41	.25-06 Remington	0.5	2.2	5.6	11.0	19.0	0.8	3.3	7.9	14.7	24.2	0.6	2.5	6.0	11.4	19.1
42	.25-06 Remington	0.5	2.5	6.4	13.4	24.9	1.2	5.1	12.4	23.8	40.3	0.6	2.6	6.6	13.1	23.2
43	.25-20 Winchester	2.5	12.2	32.1	64.9	113.1	4.0	15.7	33.6	56.7	84.4	2.8	12.7	32.1	63.3	108.4
44	.25-35 Winchester	1.0	4.8	13.0	27.7	51.8	1.9	8.4	20.4	38.7	63.1	1.1	5.1	13.0	26.5	47.3
45	.25-35 Winchester	1.0	5.1	13.9	30.4	57.6	2.2	9.7	23.6	44.7	72.0	1.2	5.3	13.7	28.4	51.4
46	.250 Savage	0.5	2.4	6.3	13.1	23.9	1.1	4.8	11.6	22.1	37.2	0.6	2.6	6.5	12.9	22.6
47	.250 Savage	0.6	2.8	7.2	14.6	26.5	1.1	4.9	11.7	22.2	37.2	0.7	3.0	7.4	14.6	25.4
48	.250 Savage	0.6	2.9	7.5	15.7	29.1	1.3	5.6	13.5	25.8	43.6	0.7	3.1	7.7	15.4	27.2
49	.256 Win. Mag.	0.7	4.1	13.0	32.3	66.9	2.8	13.1	34.0	65.8	105.8	0.8	4.0	11.6	26.9	53.3
50	.256 Roberts	0.5	2.2	5.7	11.8	21.5	1.1	4.5	10.8	20.6	34.7	0.6	2.4	6.0	11.7	20.5
51	.257 Roberts	0.6	2.7	7.1	14.7	27.2	1.2	5.3	12.9	24.7	41.7	0.7	2.9	7.3	14.4	25.5
52	.257 Roberts	0.7	3.3	8.9	18.8	35.4	1.5	6.5	15.8	30.5	51.4	0.8	3.6	9.0	18.1	32.5
53	.264 Win. Mag.	0.4	2.0	5.3	10.9	19.9	1.0	4.4	10.6	20.2	34.0	0.5	2.2	5.5	10.8	18.9
54	.264 Win. Mag.	0.5	2.3	5.5	10.8	18.7	0.8	3.2	7.4	13.8	22.6	0.6	2.5	6.0	11.3	19.0
55	.270 Winchester	0.4	1.9	4.8	9.9	18.1	1.0	4.2	10.1	19.2	32.2	0.5	2.0	5.0	9.8	17.1
56	.270 Winchester	0.5	2.1	5.3	10.4	18.0	0.6	3.2	7.4	13.9	22.7	0.6	2.4	5.7	10.8	18.2
57	.270 Winchester	0.5	2.2	5.5	10.8	19.0	0.8	3.5	8.3	15.6	25.8	0.6	2.4	5.8	11.2	18.9
58	.270 Winchester	0.6	2.5	6.3	12.5	21.9	0.9	3.8	9.0	16.8	27.8	0.6	2.8	6.7	12.8	21.8
59	.270 Winchester	0.6	2.7	7.0	14.5	26.6	1.2	5.2	12.5	23.9	40.2	0.7	2.9	7.2	14.3	25.1
60	.270 Winchester	0.7	2.9	7.3	14.6	25.9	1.0	4.3	10.2	19.2	31.7	0.7	3.2	7.8	15.0	25.5
61	.280 Remington	0.5	2.3	5.7	11.1	19.0	0.7	3.1	7.3	13.5	22.0	0.6	2.6	6.1	11.6	19.4
62	.280 Remington	0.6	2.8	7.1	14.5	26.1	1.1	4.8	11.4	21.7	36.3	0.7	3.0	7.4	14.5	25.1
63	.284 Winchester	0.5	2.2	5.5	11.0	19.5	0.9	3.8	9.0	16.9	28.1	0.6	2.4	5.8	11.2	19.2
64	.284 Winchester	0.6	2.6	6.5	12.9	22.6	0.9	3.9	9.2	17.2	28.4	0.7	2.8	6.9	13.2	22.4
65	.30 Carbine	1.4	7.2	20.9	46.2	86.2	3.4	15.0	35.5	63.2	96.7	1.5	7.3	19.7	41.6	75.4
66	.30 Remington	1.1	5.3	14.1	29.7	54.8	2.0	8.4	20.3	38.3	61.8	1.3	5.6	14.2	28.6	50.6
67	.30-30 Winchester	0.9	4.3	11.8	25.7	49.1	2.0	8.5	20.9	40.1	66.1	1.0	4.5	11.7	24.1	43.9
68	.30-30 Winchester	0.9	4.5	12.8	28.5	55.4	2.2	9.8	24.4	46.7	76.0	1.0	4.7	12.4	26.1	48.3
69	.30-30 Winchester	1.0	4.9	13.0	27.4	50.8	1.9	8.0	19.4	36.7	59.8	1.2	5.2	13.1	26.4	46.9
70	.300 H&H Magnum	0.5	2.2	5.5	11.0	19.5	0.9	3.8	8.9	16.8	27.9	0.6	2.4	5.8	11.3	19.2
71	.300 H&H Magnum	0.6	2.5	6.2	12.1	21.0	0.8	3.4	8.0	14.9	24.5	0.7	2.8	6.6	12.6	21.2
72	.300 H&H Magnum	0.7	3.2	8.0	15.8	27.8	1.0	4.3	10.2	19.1	31.4	0.8	3.5	8.5	16.3	27.6

73	.300 Win. Mag.	0.4	2.0	5.1	10.2	18.2	0.9	3.8	9.0	16.9	28.2	0.5	2.2	5.3	10.4	17.8
74	.300 Win. Mag.	0.6	2.5	6.2	12.1	21.0	0.8	3.4	8.0	14.9	24.5	0.7	2.8	6.6	12.6	21.2
75	.300 Win. Mag.	0.7	2.9	7.2	14.2	24.7	0.9	3.8	9.0	16.8	27.7	0.8	3.2	7.7	14.7	24.8
76	.300 Win. Mag.	0.7	3.1	7.9	16.1	29.1	1.2	5.1	12.2	23.1	38.5	0.8	3.3	8.2	16.1	27.9
77	.30-06 Springfield	0.5	2.2	6.0	13.2	26.1	1.4	6.2	15.2	30.0	52.1	0.5	2.3	5.9	12.3	22.9
78	.30-06 Springfield	0.5	2.3	5.8	11.9	21.7	1.1	4.5	10.8	20.5	34.4	0.6	2.4	6.0	11.9	20.7
79	.30-06 Springfield	0.6	2.5	6.1	12.1	21.0	0.8	3.5	8.4	15.6	25.7	0.6	2.7	6.6	12.5	21.1
80	.30-06 Springfield	0.6	2.7	6.7	13.2	23.2	0.9	3.9	9.2	17.2	28.4	0.7	2.9	7.1	13.6	23.1

EXTERIOR BALLISTICS TABLES—RIFLE

Height of Trajectory (Inches) Above or Below Sight Line
for Sights 0.9 Inches Above Bore Line

		Short Range						Long Range						
Line No.	*Cartridge*	*50*	*100*	*150*	*200*	*250*	*300*	*100*	*150*	*200*	*250*	*300*	*400*	*500*
		Yards						*Yards*						
41	.25-06 Remington	0.5	0.8	0	–2.0	–5.4	–10.2	1.8	1.5	0	–2.9	–7.2	–21.9	–42.9
42	.25-06 Remington	0.5	0.9	0	–2.3	–6.2	–12.0	2.0	1.7	0	–3.3	–8.5	–25.9	–55.5
43	.25-20 Winchester	0	–4.1	–14.4	–31.8	–57.3	–92.0	0	–8.2	–23.5	–47.0	–79.6	–175.9	–319.4
44	.25-35 Winchester	0.6	0	–3.0	–8.8	–17.9	–31.0	2.0	0	–4.8	–13.0	–25.1	–64.2	–128.7
45	.25-35 Winchester	0.6	0	–3.1	–9.2	–19.0	–33.1	2.1	0	–5.1	–13.8	–27.0	–70.1	–142.0
46	.250 Savage	0.5	0.9	0	–2.3	–5.1	–11.8	2.0	1.7	0	–3.3	–8.4	–25.2	–53.4
47	.250 Savage	0.2	0	–1.6	–4.7	–9.6	–16.5	2.3	2.0	0	–3.7	–9.5	–28.3	–59.5
48	.250 Savage	0.2	0	–1.6	–4.9	–10.0	–17.4	2.4	2.0	0	–3.9	–10.1	–30.5	–65.2
49	.256 Win. Mag.	0.3	0	–2.3	–7.3	–15.9	–29.6	1.5	0	–4.2	–12.1	–25.0	–72.1	–157.2
50	.257 Roberts	0.4	0.8	0	–2.0	–5.5	–10.6	1.8	1.5	0	–3.0	–7.5	–22.7	–48.00
51	.257 Roberts	0.6	1.0	0	–2.5	–5.9	–13.2	2.3	1.9	0	–3.7	–9.4	–28.6	–60.9
52	.257 Roberts	0.3	0	–1.9	–5.8	–11.9	–20.7	2.9	2.4	0	–4.7	–12.0	–36.7	–79.2
53	.264 Win. Mag.	0.2	0.5	0	–1.5	–4.1	–7.9	2.1	2.4	1.8	0	–3.0	–13.6	–31.9
54	.264 Win. Mag.	0.4	0.7	0	–1.9	–4.9	–9.4	2.7	3.0	2.1	0	–3.5	–15.0	–33.7
55	.270 Winchester	0.3	0.6	0	–1.6	–4.5	–8.7	2.4	2.7	1.9	0	–3.3	–15.0	–35.2
56	.270 Winchester	0.4	0.7	0	–1.9	–5.1	–9.7	1.7	1.4	0	–2.7	–6.8	–19.9	–40.5
57	.270 Winchester	0.4	0.8	0	–2.0	–5.3	–10.0	1.7	1.5	0	–2.8	–7.1	–20.8	–42.7
58	.270 Winchester	0.6	0.9	0	–2.3	–6.1	–11.7	2.1	2.7	0	–3.3	–8.2	–24.1	–49.4
59	.270 Winchester	0.6	1.0	0	–2.5	–6.8	–13.1	2.2	1.9	0	–3.6	–9.3	–28.1	–59.7
60	.270 Winchester	0.2	0	–1.7	–4.9	–10.0	–17.1	2.5	2.0	0	–3.8	–9.7	–28.4	–58.5

Height of Trajectory (Inches) Above or Below Sight Line
for Sights 0.9 Inches Above Bore Line

Line No.	Cartridge	Short Range						Long Range						
		50	100	150	200	250	300	100	150	200	250	300	400	500
					Yards						Yards			
61	.280 Remington	0.5	0.8	0	-2.1	-5.5	-10.5	1.9	1.6	0	-2.9	-7.3	-21.2	-42.9
62	.280 Remington	0.2	0	-1.5	-4.6	-9.5	-16.4	2.3	1.9	0	-3.7	-9.4	-28.1	-58.8
63	.284 Winchester	0.4	0.8	0	-2.0	-5.3	-10.1	1.7	1.5	0	-2.8	-7.2	-21.1	-43.7
64	.284 Winchester	0.6	1.0	0	-2.4	-6.3	-12.1	2.1	1.8	0	-3.4	-8.5	-24.8	-51.0
65	.30 Carbine	0.9	0	-4.5	-13.5	-28.3	-49.9	0	-4.5	-13.5	-28.3	-40.9	-118.6	-228.1
66	.30 Remington	0.7	0	-3.3	-9.7	-19.6	-33.8	2.2	0	-5.3	-14.1	-27.2	-69.0	-136.9
67	.30-30 Winchester	0.5	0	-2.6	-7.7	-16.0	-27.9	1.7	0	-4.3	-11.6	-22.7	-59.1	-120.5
68	.30-30 Winchester	0.5	0	-2.7	-8.2	-17.0	-30.0	1.8	0	-4.6	-12.5	-24.6	-65.3	-134.9
69	.30-30 Winchester	0.6	0	-3.0	-8.9	-18.0	-31.1	2.0	0	-4.8	-13.0	-25.1	-63.6	-126.7
70	.300 H&H Magnum	0.4	0.8	0	-2.0	-5.3	-10.1	1.7	1.5	0	-2.8	-7.2	-21.2	-43.8
71	.300 H&H Magnum	0.6	0.9	0	-2.3	-6.0	-11.5	2.1	1.7	0	-3.2	-8.0	-23.3	-47.4
72	.300 H&H Magnum	0.3	0	-1.9	-5.5	-11.0	-18.7	2.7	2.2	0	-4.2	-10.5	-30.7	-63.0
73	.300 Win. Mag.	0.3	0.7	0	-1.8	-4.8	-9.3	2.6	2.9	2.1	0	-3.5	-15.4	-35.5
74	.300 Win. Mag.	0.6	0.7	0	-2.3	-6.0	-11.5	2.1	1.7	0	-3.2	-8.0	-23.3	-47.4
75	.300 Win. Mag.	0.2	0	-1.7	-4.9	-9.9	-16.9	2.5	2.0	0	-3.8	-9.5	-27.5	-56.1
76	.300 Win. Mag.	0.2	0	-1.8	-5.3	-10.7	-18.4	2.6	2.2	0	-4.1	-10.5	-31.3	-65.6
77	.30-06 Springfield	0.4	0.7	0	-2.0	-5.6	-11.1	1.7	1.5	0	-3.1	-8.0	-25.5	-57.4
78	.30-06 Springfield	0.4	0.8	0	-2.1	-5.6	-10.7	1.8	1.5	0	-3.0	-7.7	-23.0	-48.5
79	.30-06 Springfield	0.6	0.9	0	-2.2	-6.0	-11.4	2.0	1.7	0	-3.2	-8.0	-23.3	-47.5
80	.30-06 Springfield	0.7	1.0	0	-2.5	-6.8	-12.5	2.2	1.8	0	-3.5	-8.8	-25.6	-52.5

EXTERIOR BALLISTICS TABLES—RIFLE

Line No.	Cartridge	Wgt. Grs.	Ballistic Coefficient	Bullet Type				Bullet Drop-Inches from Bore Line				
				I.V.I.	Federal	Rem-Peters	W-W	100	200	300	400	500
										Yards		
81	.30-06 Springfield	150	0.314	PSP,ST	HSSP	PSPCL	ST	2.20	9.51	23.18	44.90	76.92
82	.30-06 Springfield	150	0.270				PP(SP)	2.21	9.68	23.98	47.28	82.65

#	Cartridge											
83	.30-06 Springfield	180	0.499				FMCBT	2.49	10.46	24.74	46.30	76.34
84	.30-06 Springfield	180	0.438				PP(SP)	2.51	10.61	25.27	47.71	79.39
85	.30-06 Springfield	180	0.412	CPE		BP		2.52	10.68	25.55	48.45	81.01
86	.30-06 Springfield	180	0.383	ST		PSPCL	ST	2.53	10.78	25.92	49.43	83.18
87	.30-06 Springfield	180	0.329		HSSP			2.56	11.01	26.82	51.88	88.71
88	.30-06 Springfield	180	0.248	KKSP		SPCL	PP(SP)	2.62	11.59	29.14	58.46	104.08
89	.30-06 Springfield	220	0.380				ST	3.19	13.64	32.93	63.11	106.73
90	.30-06 Springfield	220	0.294	KKSP		SPCL	PP(SP)	3.25	14.21	35.17	69.27	120.65
91	.30-40 Krag	180	0.383			PSPCL	ST	3.13	13.39	32.31	61.85	104.49
92	.30-40 Krag	180	0.248			SPCL	P(SP)	3.25	14.46	36.56	73.74	131.64
93	.30-40 Krag	220	0.382				ST	3.98	17.08	41.38	79.53	134.80
94	.300 Savage	150	0.314	PSP,ST		PSPCL	ST	2.71	11.71	28.69	55.83	96.11
95	.300 Savage	150	0.270				PP(ST)	2.74	12.03	29.95	59.41	104.42
96	.300 Savage	150	0.223			SPCL	ST	2.79	12.55	32.07	65.58	119.05
97	.300 Savage	180	0.383	ST	HSSP	PSPCL		3.35	14.35	34.66	66.43	112.34
98	.300 Savage	180	0.248	KKSP		SPCL	PP(SP)	3.48	15.51	39.29	79.35	141.62
99	.303 Savage	180	0.246			SPCL		4.30	19.27	49.05	99.23	176.37
100	.303 Savage	190	0.252	KKSP	HSSP		ST	5.14	23.05	58.58	117.83	207.22
101	.303 British	150	0.300	PSP,ST				2.57	11.18	27.48	53.72	92.93
102	.303 British	180	0.407	CPE				3.04	12.95	21.10	59.21	99.43
103	.303 British	180	0.369	ST	HSSP		PP(SP)	3.06	13.12	31.75	60.96	103.31
104	.303 British	180	0.246					3.17	14.11	35.68	71.99	128.31
105	.303 British	180	0.237	KKSP		SPCL		3.18	14.24	36.27	73.52	132.18
106	.303 British	215	0.285	KKSP		SP		4.04	17.78	44.35	87.95	153.84
107	.308 Winchester	110	0.186				PSP	1.94	8.83	22.97	48.08	90.05
108	.308 Winchester	125	0.268				PSP	2.03	8.86	21.93	43.19	75.42
109	.308 Winchester	150	0.314	PSP,ST	HSSP	PSPCL	ST	2.25	10.14	24.76	48.03	82.41
110	.308 Winchester	150	0.270				PP(SP)	2.38	10.41	25.82	51.00	89.34
111	.308 Winchester	180	0.383	ST	HSSP	PSPCL	ST	2.69	11.47	27.60	52.69	88.77
112	.308 Winchester	180	0.248			SPCL	PP(SP)	2.78	12.35	31.09	62.47	111.35
113	.308 Winchester	180	0.237	KKSP				2.80	12.47	31.58	63.91	114.76
114	.308 Winchester	200	0.345	KKSP			ST	3.10	13.37	32.54	62.89	107.38
115	.32 Remington	170	0.237			SPCL		4.23	19.04	48.68	98.95	176.61
116	.32 Remington	170	0.213	KKSP,ST	HSSP	HPCL,SPCL	ST	4.29	19.63	51.12	105.70	190.93
117	.32 Win. Spl.	170	0.239				PP(SP),ST	3.82	17.14	43.75	88.95	159.33
118	.32 Win. Spl.	170	0.205					3.89	17.85	46.71	97.33	177.55
119	.32-20 Winchester	100	0.166			SP, Lead	SP, Lead	13.53	60.27	148.69	287.13	484.51
120	.32-20 Winchester	115	0.200	SP				8.99	41.17	104.53	206.30	353.32

EXTERIOR BALLISTICS TABLES—RIFLE

Line No.	Cartridge	Velocity in fps at						Energy in ft.-lbs. at						Time of Flight in Seconds				
		0	100	200	300	400	500	0	100	200	300	400	500	100	200	300	400	500
				Yards						Yards						Yards		
81	.30-06 Springfield	2910	2617	2342	2083	1843	1622	2820	2281	1827	1445	1131	876	0.109	0.230	0.366	0.519	0.693
82	.30-06 Springfield	2920	2580	2265	1972	1704	1466	2839	2217	1708	1295	967	716	0.109	0.233	0.375	0.539	0.729
83	.30-06 Springfield	2700	2522	2350	2185	2027	1876	2913	2542	2207	1908	1642	1406	0.115	0.238	0.371	0.513	0.667
84	.30-06 Springfield	2700	2497	2303	2118	1942	1775	2913	2492	2119	1793	1507	1259	0.116	0.241	0.376	0.524	0.686
85	.30-06 Springfield	2700	2485	2280	2084	1899	1725	2913	2468	2077	1736	1441	1189	0.116	0.242	0.380	0.530	0.696
86	.30-06 Springfield	2700	2469	2250	2042	1846	1663	2913	2436	2023	1666	1362	1105	0.116	0.243	0.383	0.538	0.709
87	.30-06 Springfield	2700	2432	2180	1943	1723	1523	2913	2364	1899	1509	1186	927	0.117	0.247	0.393	0.557	0.742
88	.30-06 Springfield	2700	2348	2023	1727	1466	1251	2913	2203	1635	1192	859	625	0.119	0.257	0.417	0.606	0.828
89	.30-06 Springfield	2410	2192	1985	1791	1611	1448	2837	2347	1924	1567	1268	1024	0.131	0.274	0.438	0.610	0.807
90	.30-06 Springfield	2410	2130	1870	1632	1422	1246	2837	2216	1708	1301	988	758	0.132	0.283	0.455	0.652	0.878
91	.30-40 Krag	2430	2213	2007	1813	1632	1468	2360	1957	1610	1314	1064	861	0.129	0.272	0.429	0.604	0.797
92	.30-40 Krag	2430	2098	1795	1525	1298	1128	2360	1759	1288	929	673	508	0.133	0.287	0.469	0.683	0.931
93	.30-40 Krag	2160	1956	1765	1587	1427	1287	2279	1869	1522	1230	995	809	0.146	0.307	0.487	0.686	0.908
94	.300 Savage	2630	2354	2095	1853	1631	1433	2303	1845	1462	1143	886	684	0.121	0.256	0.408	0.581	0.777
95	.300 Savage	2630	2311	2015	1743	1500	1295	2303	1779	1352	1012	749	558	0.122	0.261	0.421	0.607	0.822
96	.300 Savage	2630	2247	1897	1585	1324	1131	2303	1681	1198	837	584	426	0.123	0.269	0.442	0.649	0.896
97	.300 Savage	2350	2137	1935	1745	1570	1413	2207	1825	1496	1217	985	798	0.134	0.281	0.445	0.626	0.828
98	.300 Savage	2350	2025	1728	1467	1252	1098	2207	1639	1193	860	626	482	0.138	0.298	0.487	0.708	0.983
99	.303 Savage	2120	1813	1539	1308	1134	1021	1796	1314	946	684	514	417	0.153	0.333	0.545	0.792	1.071
100	.303 Savage	1940	1657	1410	1211	1072	982	1588	1158	839	619	485	407	0.167	0.364	0.594	0.858	1.152
101	.303 British	2700	2407	2132	1876	1642	1434	2428	1929	1514	1172	898	685	0.118	0.250	0.400	0.571	0.767
102	.303 British	2460	2254	2057	1871	1697	1536	2418	2030	1691	1399	1151	943	0.117	0.267	0.420	0.588	0.774
103	.303 British	2460	2233	2018	1816	1629	1459	2418	1993	1627	1318	1060	851	0.128	0.269	0.426	0.600	0.795
104	.303 British	2460	2124	1817	1542	1311	1137	2418	1803	1319	950	687	517	0.131	0.284	0.463	0.675	0.921
105	.303 British	2460	2112	1794	1512	1279	1109	2418	1782	1286	914	654	491	0.132	0.286	0.468	0.684	0.937
106	.303 British	2170	1899	1652	1432	1250	1113	2240	1721	1303	979	746	591	0.148	0.317	0.512	0.737	0.992
107	.308 Winchester	3180	2666	2206	1795	1444	1178	2470	1736	1188	787	509	339	0.103	0.227	0.378	0.564	0.795
108	.308 Winchester	3050	2697	2370	2067	1788	1537	2582	2019	1559	1186	887	656	0.105	0.224	0.359	0.515	0.696
109	.308 Winchester	2820	2532	2263	2009	1774	1560	2648	2135	1705	1344	1048	810	0.112	0.238	0.378	0.537	0.718
110	.308 Winchester	2820	2488	2179	1893	1633	1405	2648	2061	1581	1193	888	657	0.113	0.242	0.390	0.561	0.759
111	.308 Winchester	2620	2393	2178	1974	1782	1604	2743	2288	1896	1557	1269	1028	0.120	0.251	0.396	0.556	0.733
112	.308 Winchester	2620	2274	1955	1666	1414	1212	2743	2066	1527	1109	799	587	0.123	0.265	0.432	0.627	0.857
113	.308 Winchester	2620	2259	1928	1630	1374	1175	2743	2039	1435	1062	754	552	0.123	0.267	0.436	0.637	0.874
114	.308 Winchester	2450	2208	1980	1767	1572	1397	2665	2165	1741	1386	1097	867	0.129	0.272	0.433	0.613	0.816
115	.32 Remington	2140	1822	1538	1301	1125	1012	1728	1253	893	639	478	387	0.152	0.331	0.544	0.793	1.075

Line No.	Cartridge																	
116	.32 Remington	2140	1785	1228	1475	1063	962	1728	1203	821	569	426	349	0.154	0.339	0.562	0.826	1.124
117	.32 Win. Spl.	2250	1921	1372	1626	1174	1044	1911	1393	998	710	520	411	0.144	0.314	0.515	0.752	1.025
118	.32 Win. Spl.	2250	1870	1267	1537	1082	971	1911	1320	892	606	442	356	0.146	0.323	0.539	0.797	1.091
119	.32-20 Winchester	1210	1021	834	913	769	712	325	231	185	154	131	113	0.272	0.584	0.928	1.303	1.709
120	.32-20 Winchester	1490	1226	951	1054	876	816	567	384	284	231	196	170	0.223	0.488	0.789	1.118	1.473

EXTERIOR BALLISTICS TABLES—RIFLE

Line No.	Cartridge	Midrange Trajectory Height in Inches from Bore Line					Drift in Inches in a 10 mph Wind					45° Angle Up- or Downhill Hold (Inches) under Target				
		100	200	300 Yards	400	500	100	200	300 Yards	400	500	100	200	300 Yards	400	500
81	.30-06 Springfield	0.6	2.5	6.4	12.9	23.0	1.0	4.2	9.9	18.7	31.2	0.6	2.8	6.8	13.2	22.5
82	.30-06 Springfield	0.6	2.6	6.8	14.0	22.5	1.2	4.9	11.8	22.6	37.9	0.6	2.8	7.0	13.8	24.2
83	.30-06 Springfield	0.6	2.7	6.6	12.7	21.4	0.7	2.8	6.6	12.1	19.6	0.7	3.1	7.2	13.6	22.4
84	.30-06 Springfield	0.6	2.8	6.8	13.2	22.6	0.8	3.2	7.6	14.1	23.0	0.7	3.1	7.4	14.0	23.3
85	.30-06 Springfield	0.6	2.8	6.9	13.5	23.3	0.8	3.5	8.1	15.1	24.7	0.7	3.1	7.5	14.2	23.7
86	.30-06 Springfield	0.7	2.9	7.1	13.9	24.2	0.9	3.7	8.8	16.5	27.1	0.7	3.2	7.6	14.5	24.4
87	.30-06 Springfield	0.7	3.0	7.4	14.9	26.5	1.1	4.4	10.5	19.8	32.9	0.7	3.2	7.9	15.2	26.0
88	.30-06 Springfield	0.7	3.2	8.4	17.6	32.9	1.4	6.1	14.8	28.5	48.0	0.8	3.4	8.5	17.1	30.5
89	.30-06 Springfield	0.8	3.6	9.1	17.9	31.3	1.1	4.5	10.6	19.7	32.4	0.9	4.0	9.6	18.5	31.3
90	.30-06 Springfield	0.8	3.9	9.9	20.4	37.1	1.4	5.9	14.3	27.1	44.9	1.0	4.2	10.3	20.3	35.3
91	.30-40 Krag	0.8	3.6	8.9	17.5	30.6	1.0	4.4	10.3	19.3	31.7	0.9	3.9	9.5	18.1	30.6
92	.30-40 Krag	0.9	4.0	10.6	22.4	41.9	1.7	7.1	17.4	33.2	55.3	1.0	4.2	10.7	21.6	38.6
93	.30-40 Krag	1.0	4.6	11.4	22.7	39.7	1.2	5.2	12.3	23.0	37.5	1.2	5.0	12.1	23.3	39.5
94	.300 Savage	0.7	3.2	8.0	16.2	29.0	1.1	4.8	11.6	21.9	36.3	0.8	3.4	8.4	16.4	28.2
95	.300 Savage	0.7	3.3	8.5	17.7	32.5	1.3	5.7	13.8	26.5	44.3	0.8	3.5	8.8	17.4	30.6
96	.300 Savage	0.7	3.5	9.4	20.2	38.7	1.6	7.2	17.5	34.0	57.3	0.8	3.7	9.4	19.2	34.9
97	.300 Savage	0.9	3.8	9.5	18.9	33.0	1.1	4.6	10.9	20.3	33.3	1.0	4.2	10.2	19.5	32.9
98	.300 Savage	0.9	4.3	11.4	24.2	45.1	1.7	7.5	18.2	34.8	57.5	1.0	4.5	11.5	23.2	41.5
99	.303 Savage	1.1	5.3	14.3	30.3	56.2	2.0	8.8	21.1	39.7	64.1	1.3	5.6	14.4	29.1	51.7
100	.303 Savage	1.4	6.4	17.1	35.9	65.3	2.3	9.6	22.9	42.2	66.6	1.5	6.8	17.2	34.5	60.7
101	.303 British	0.7	3.0	7.7	15.7	28.2	1.2	4.9	11.8	22.3	37.2	0.8	3.3	8.0	15.7	27.2
102	.303 British	0.8	3.4	8.5	16.7	28.8	1.0	4.0	9.5	17.6	28.9	0.9	3.8	9.1	17.3	29.1
103	.303 British	0.8	3.5	8.7	17.4	30.4	1.1	4.5	10.6	19.8	32.6	0.9	3.8	9.3	17.9	30.3

EXTERIOR BALLISTICS TABLES—RIFLE (continued)

Line No.	Cartridge	Midrange Trajectory Height in Inches from Bore Line					Drift in Inches in a 10 mph Wind					45° Angle Up- or Downhill Hold (Inches) under Target				
		100	200	300 Yards	400	500	100	200	300 Yards	400	500	100	200	300 Yards	400	500
104	.303 British	0.8	3.9	10.3	21.9	41.0	1.6	7.1	17.2	32.9	54.8	0.9	4.1	10.5	21.1	37.7
105	.303 British	0.8	3.9	10.5	22.5	42.5	1.7	7.4	18.0	34.6	57.7	0.9	4.2	10.6	21.5	38.7
106	.303 British	1.1	4.9	12.7	26.2	47.7	1.7	7.2	17.2	32.4	53.0	1.2	5.2	13.0	25.8	45.1
107	.308 Winchester	0.5	2.5	6.8	15.2	30.2	1.5	6.7	16.6	32.9	57.0	0.6	2.6	6.7	14.1	26.4
108	.308 Winchester	0.5	2.4	6.2	12.7	23.2	1.1	4.7	11.2	21.4	35.9	0.6	2.6	6.4	12.7	22.1
109	.308 Winchester	0.6	2.7	6.9	13.9	24.7	1.0	4.4	10.4	19.7	32.7	0.7	3.0	7.3	14.1	24.1
110	.308 Winchester	0.6	2.8	7.3	15.1	27.6	1.2	5.2	12.5	23.8	40.0	0.7	3.0	7.6	14.9	26.2
111	.308 Winchester	0.7	3.0	7.6	14.9	25.9	0.9	3.9	9.2	17.2	28.3	0.8	3.4	8.1	15.4	26.0
112	.308 Winchester	0.7	3.4	9.0	18.9	35.3	1.5	6.4	15.5	29.8	50.1	0.8	3.6	9.1	18.3	32.6
113	.308 Winchester	0.7	3.4	9.2	19.5	36.7	1.6	6.7	16.3	31.5	53.1	0.8	3.7	9.2	18.7	33.6
114	.308 Winchester	0.8	3.6	9.0	18.1	32.0	1.2	4.9	11.5	21.7	35.8	0.9	3.9	9.5	18.4	31.5
115	.32 Remington	1.1	5.3	14.3	30.4	56.6	2.1	9.0	21.7	40.8	65.8	1.2	5.6	14.3	29.0	51.7
116	.32 Remington	1.1	5.5	15.3	33.2	62.5	2.4	10.3	24.9	46.7	74.4	1.3	5.8	15.0	31.0	55.9
117	.32 Win. Spl.	1.0	4.8	12.8	27.3	51.2	1.9	8.4	20.6	38.6	63.0	1.1	5.0	12.8	26.1	46.7
118	.32 Win. Spl.	1.0	5.0	14.0	30.8	58.7	2.3	10.0	24.5	46.4	74.6	1.1	5.2	13.7	28.5	52.0
119	.32-20 Winchester	3.6	16.6	42.1	83.3	143.4	4.3	15.5	32.4	54.8	82.7	4.0	17.7	43.6	84.1	141.9
120	.32-20 Winchester	2.4	11.6	30.6	62.0	108.2	3.8	15.1	32.5	55.0	82.1	2.6	12.1	30.6	60.4	103.5

EXTERIOR BALLISTICS TABLES—RIFLE

Height of Trajectory (Inches) Above or Below Sight Line
for Sights 0.9 Inches Above Bore Line

Line No.	Cartridge	Short Range						Long Range						
		50	100	150 Yards	200	250	300	100	150	200	250 Yards	300	400	500
81	.30-06 Springfield	0.6	0.9	0	-2.3	-6.3	-12.0	2.1	1.8	0	-3.3	-8.5	-25.0	-51.8
82	.30-06 Springfield	0.6	1.0	0	-2.4	-6.6	-12.7	2.2	1.8	0	-3.5	-9.0	-27.0	-57.1
83	.30-06 Springfield	0.2	0	-1.6	-4.6	-9.2	-15.5	2.3	1.9	0	-3.5	-8.6	-24.5	-48.8

84	.30-06 Springfield	0.2	0	-1.6	-4.7	-9.4	-15.9	2.3	1.9	0	-3.6	-8.9	-25.6	-51.5
85	.30-06 Springfield	0.2	0	-1.6	-4.7	-9.6	-16.2	2.4	2.0	0	-3.6	-9.1	-26.2	-53.0
86	.30-06 Springfield	0.2	0	-1.6	-4.8	-9.7	-16.5	2.4	2.0	0	-3.7	-9.3	-27.0	-54.9
87	.30-06 Springfield	0.2	0	-1.7	-5.0	-10.2	-17.4	2.5	2.1	0	-3.9	-9.9	-29.0	-59.8
88	.30-06 Springfield	0.2	0	-1.8	-5.5	-11.2	-19.5	2.7	2.3	0	-4.4	-11.3	-34.4	-73.7
89	.30-06 Springfield	0.4	0	-2.2	-6.4	-12.7	-21.6	1.5	0	-3.5	-9.1	-17.2	-41.8	-79.9
90	.30-06 Springfield	0.4	0	-2.3	-6.8	-13.8	-23.6	1.5	0	-3.7	-9.9	-19.0	-47.4	-93.1
91	.30-40 Krag	0.4	0	-2.1	-6.2	-12.5	-21.1	1.4	0	-3.4	-8.9	-16.8	-40.9	-78.1
92	.30-40 Krag	0.4	0	-2.4	-7.1	-14.5	-25.0	1.6	0	-3.9	-10.5	-20.3	-51.7	-103.9
93	.30-40 Krag	0.6	0	-2.9	-8.2	-16.4	-27.6	1.9	0	-4.4	-11.6	-21.9	-52.3	-101.8
94	.300 Savage	0.3	0	-1.8	-5.4	-11.0	-18.8	2.7	2.2	0	-4.2	-10.7	-31.5	-65.5
95	.300 Savage	0.3	0	-1.9	-5.7	-11.6	-19.9	2.8	2.3	0	-4.5	-11.5	-34.4	-73.0
96	.300 Savage	0.3	0	-2.0	-6.1	-12.5	-21.9	1.3	0	-3.4	-9.2	-17.9	-46.3	-94.8
97	.300 Savage	0.4	0	-2.3	-6.7	-13.5	-22.8	1.5	0	-3.6	-9.8	-18.2	-44.1	-84.2
98	.300 Savage	0.5	0	-2.6	-7.7	-15.6	-27.1	1.7	0	-4.2	-11.3	-21.9	-55.8	-112.0
99	.303 Savage	0.7	0	-3.3	-9.8	-19.9	-34.4	2.2	0	-5.3	-14.4	-27.7	-70.5	-140.2
100	.303 Savage	0.9	0	-4.1	-11.9	-24.1	-41.4	2.7	0	-6.4	-17.3	-33.2	-83.7	-164.3
101	.303 British	0.2	0	-1.7	-5.1	-10.5	-18.0	2.6	2.1	0	-4.0	-10.3	-30.5	-63.6
102	.303 British	0.3	0	-1.9	-5.6	-11.3	-19.1	2.8	2.3	0	-4.2	-10.6	-30.7	-62.3
103	.303 British	0.3	0	-2.0	-5.8	-11.6	-19.6	2.9	2.4	0	-4.4	-11.0	-32.0	-65.5
104	.303 British	0.3	0	-2.2	-6.5	-13.3	-23.0	1.4	0	-3.6	-9.7	-18.7	-47.7	-96.0
105	.303 British	0.3	0	-2.2	-6.6	-13.5	-23.5	1.5	0	-3.6	-9.8	-19.1	-49.0	-99.2
106	.303 British	0.6	0	-3.0	-8.8	-17.8	-30.4	2.0	0	-4.8	-12.8	-24.4	-61.1	-120.0
107	.308 Winchester	0.5	0.9	0	-2.3	-6.5	-12.8	2.0	1.8	0	-3.5	-9.3	-29.5	-66.7
108	.308 Winchester	0.5	0.8	0	-2.2	-6.0	-11.5	2.0	1.7	0	-3.2	-8.2	-24.6	-51.9
109	.308 Winchester	0.2	0	-1.5	-4.5	-9.3	-15.9	2.3	1.9	0	-3.6	-9.1	-26.9	-55.7
110	.308 Winchester	0.2	0	-1.6	-4.8	-9.8	-16.9	2.4	2.0	0	-3.8	-9.8	-29.3	-62.0
111	.308 Winchester	0.2	0	-1.8	-5.2	-10.4	-17.7	2.6	2.1	0	-4.0	-9.9	-28.9	-58.8
112	.308 Winchester	0.3	0	-2.0	-5.9	-12.1	-20.9	2.9	2.4	0	-4.7	-12.1	-36.9	-79.1
113	.308 Winchester	0.3	0	-2.0	-6.0	-12.3	-21.4	3.0	2.5	0	-4.8	-12.4	-38.1	-82.2
114	.308 Winchester	0.4	0	-2.1	-6.3	-12.6	-21.4	1.4	0	-3.4	-9.0	-17.2	-42.1	-81.1
115	.32 Remington	0.7	0	-3.3	-9.7	-19.8	-34.2	2.2	0	-5.3	-14.3	-27.6	-70.6	-140.9
116	.32 Remington	0.7	0	-3.4	-10.2	-20.9	-36.5	2.3	0	-5.6	-15.2	-29.6	-76.7	-154.5
117	.32 Win. Spl.	0.6	0	-2.9	-8.6	-17.6	-30.5	1.9	0	-4.7	-12.7	-24.7	-63.2	-126.9
118	.32 Win. Spl.	0.6	0	-3.1	-9.2	-19.0	-33.2	2.0	0	-5.1	-13.8	-27.1	-70.9	-144.3
119	.32-20 Winchester	0	-6.3	-20.9	-44.9	-79.3	-125.1	0	-11.5	-32.3	-63.3	-106.3	-230.3	-413.3
120	.32-30 Winchester	0	-3.9	-13.6	-30.1	-54.4	-87.5	0	-7.7	-22.3	-44.7	-75.8	-167.7	-304.8

EXTERIOR BALLISTICS TABLES—RIFLE

Line No.	Cartridge	Wgt. Grs.	Ballistic Coefficient	Bullet Type				Bullet Drop-Inches from Bore Line				
				I.V.I.	Federal	Rem-Peters	W-W	100	200	300 Yards	400	500
121	.32-40 Winchester	170	0.279	KKSP				8.20	36.27	89.98	175.06	296.71
122	.338 Win. Mag.	200	0.308				PP(SP)	2.13	9.21	22.48	43.60	74.81
123	.338 Win. Mag.	250	0.329				ST,PP,(SP)	2.63	11.36	27.67	53.57	91.65
124	.338 Win. Mag.	300	0.297				PP(SP)	3.19	13.94	34.45	67.74	117.79
125	.348 Winchester	200	0.276				ST	2.98	13.10	32.61	64.64	113.44
126	.348 Winchester	200	0.225			SPCL		3.04	13.70	35.06	71.73	130.04
127	.35 Remington	150	0.184			PSPCL		3.77	17.61	47.02	99.90	184.28
128	.35 Remington	200	0.193		HSSP	SPCL	PP(SP),(ST)	4.89	22.82	60.42	125.94	227.21
129	.350 Rem. Mag.	200	0.294	SP		PSPCL		2.56	11.13	27.42	53.71	93.15
130	.350 Rem. Mag.	250	0.375			PSPCL		3.22	13.78	33.32	63.93	108.26
131	.351 Win. S.L.	180	0.233			SP	SP	5.71	25.91	66.47	134.30	236.03
132	.358 Winchester	200	0.261				ST	3.07	13.58	34.05	68.04	120.42
133	.358 Winchester	250	0.326				ST	3.77	16.39	40.26	78.53	135.25
134	.375 H&H Magnum	270	0.326			SP	PP(SP)	2.58	11.12	27.10	52.49	89.88
135	.375 H&H Magnum	300	0.324				ST	2.92	12.63	30.88	60.00	103.05
136	.375 H&H Magnum	300	0.234			MC	FMC	3.01	13.46	34.26	69.63	125.45
137	.38-40 Winchester	180	0.171			SP	SP	14.48	63.73	155.84	298.91	501.66
138	.38-55 Winchester	255	0.311	SP				7.51	32.95	81.28	157.79	267.43
139	.44 Rem. Mag.	240	0.163	SP				6.66	32.00	85.57	176.47	313.15
140	.44 Rem. Mag.	240	0.166			SP		6.64	31.79	84.81	174.71	309.75
141	.44 Rem. Mag.	240	0.158		HSP	SP	HSP	6.70	32.41	87.00	179.83	319.60
142	.44-40 Winchester	200	0.160	SP		SP	SP	13.97	62.16	153.25	295.94	499.62
143	.444 Marlin	240	0.146			SP		3.75	18.34	51.57	114.05	215.76
144	.45-70 Govt.	405	0.280			SP	SP	10.77	47.01	114.43	218.23	363.32
145	.458 Win. Mag.	500	0.345			MC	FMC	4.50	19.51	47.79	92.83	158.92
146	.458 Win. Mag.	510	0.274			SP	SP	4.59	20.38	51.17	102.01	178.78

EXTERIOR BALLISTICS TABLES—RIFLE

Line No.	Cartridge	Velocity in fps at						Energy in ft.-lbs. at						Time of Flight in Seconds				
		0	100	200	300	400	500	0	100	200	300	400	500	100	200	300	400	500
				Yards						Yards						Yards		
121	.32-40 Winchester	1530	1325	1164	1052	974	915	883	663	511	418	358	316	0.211	0.453	0.725	1.022	1.340
122	.338 Win. Mag.	2960	2658	2375	2109	1862	1635	3890	3137	2505	1975	1539	1187	0.107	0.226	0.360	0.512	0.684
123	.338 Win. Mag.	2660	2395	2145	1910	1693	1497	3927	3184	2554	2025	1591	1244	0.119	0.251	0.399	0.566	0.755
124	.338 Win. Mag.	2430	2152	1893	1655	1443	1265	3933	3084	2387	1824	1387	1066	0.131	0.280	0.449	0.644	0.866
125	.348 Winchester	2520	2215	1931	1672	1443	1253	2820	2178	1656	1241	925	697	0.127	0.272	0.439	0.632	0.856
126	.348 Winchester	2520	2149	1812	1514	1270	1097	2820	2051	1458	1018	726	534	0.129	0.281	0.462	0.679	0.935
127	.35 Remington	2300	1874	1506	1218	1039	934	1762	1169	755	494	359	291	0.145	0.323	0.546	0.814	1.120
128	.35 Remington	2020	1645	1335	1114	985	901	1812	1203	791	551	431	360	0.165	0.368	0.615	0.903	1.222
129	.350 Rem. Mag.	2710	2410	2130	1870	1631	1421	3261	2579	2014	1553	1181	897	0.117	0.250	0.400	0.572	0.769
130	.350 Rem. Mag.	2400	2180	1971	1776	1594	1431	3197	2638	2156	1751	1410	1137	0.131	0.276	0.436	0.615	0.813
131	.351 Win. S.L.	1850	1556	1310	1128	1012	933	1368	968	686	508	409	348	0.177	0.387	0.635	0.917	1.227
132	.358 Winchester	2490	2171	1876	1610	1379	1194	2753	2093	1563	1151	844	633	0.129	0.278	0.450	0.652	0.887
133	.358 Winchester	2230	1988	1762	1557	1375	1224	2760	2194	1723	1346	1049	832	0.143	0.303	0.484	0.689	0.921
134	.375 H&H Magnum	2690	2420	2166	1928	1707	1506	4337	3510	2812	2228	1747	1360	0.118	0.249	0.395	0.561	0.748
135	.375 H&H Magnum	2530	2268	2022	1793	1583	1397	4263	3426	2723	2141	1669	1300	0.125	0.265	0.423	0.601	0.803
136	.375 H&H Magnum	2530	2171	1843	1551	1307	1126	4263	3139	2262	1602	1138	844	0.128	0.278	0.456	0.667	0.915
137	.38-40 Winchester	1160	999	901	827	764	710	538	399	324	273	233	201	0.281	0.598	0.946	1.324	1.732
138	.38-55 Winchester	1590	1395	1233	1110	1023	960	1431	1102	861	698	592	522	0.202	0.431	0.688	0.970	1.279
139	.44 Rem. Mag.	1760	1374	1107	964	872	799	1650	1006	653	495	405	340	0.193	0.439	0.731	1.059	1.419
140	.44 Rem. Mag.	1760	1380	1114	970	878	805	1650	1015	661	501	411	345	0.193	0.437	0.727	1.053	1.410
141	.44 Rem. Mag.	1760	1362	1094	953	861	788	1650	988	638	484	395	331	0.194	0.442	0.738	1.070	1.434
142	.44-40 Winchester	1190	1006	900	822	756	699	629	449	360	300	254	217	0.281	0.598	0.946	1.324	1.732
143	.444 Marlin	2350	1815	1377	1087	941	845	2942	1755	1010	630	472	380	0.145	0.336	0.583	0.882	1.219
144	.45-70 Govt.	1330	1168	1055	977	917	868	1590	1227	1001	858	756	677	0.241	0.512	0.809	1.126	1.462
145	.458 Win. Mag.	2040	1823	1623	1442	1287	1161	4620	3689	2924	2308	1839	1496	0.156	0.330	0.526	0.747	0.993
146	.458 Win. Mag.	2040	1770	1527	1319	1157	1046	4712	3547	2640	1970	1516	1239	0.158	0.341	0.552	0.796	1.069

EXTERIOR BALLISTICS TABLES—RIFLE

Line No.	Cartridge	Midrange Trajectory Height in Inches from Bore Line					Drift in Inches in a 10 mph Wind					45° Angle Up- or Downhill Hold (Inches) under Target				
		100	200	300 Yards	400	500	100	200	300 Yards	400	500	100	200	300 Yards	400	500
121	.32-40 Winchester	2.1	9.9	25.6	51.3	88.8	2.6	10.7	24.1	41.8	63.3	2.4	10.6	26.4	51.3	86.9
122	.338 Win. Mag.	0.6	2.5	6.3	12.6	22.4	1.0	4.2	9.9	18.7	31.2	0.6	2.7	6.6	12.8	21.9
123	.338 Win. Mag.	0.7	3.0	7.7	15.4	27.4	1.1	4.5	10.8	20.3	33.6	0.6	3.3	8.1	15.7	26.8
124	.338 Win. Mag.	0.8	3.8	9.7	19.9	36.1	1.4	5.8	13.9	26.4	43.8	0.9	4.1	10.1	19.8	34.5
125	.348 Winchester	0.8	3.6	9.3	19.2	35.2	1.4	6.0	14.4	27.5	45.9	0.9	3.8	9.6	18.9	33.2
126	.348 Winchester	0.8	3.8	10.3	22.2	42.2	1.7	7.6	18.5	35.7	59.7	0.9	4.0	10.3	21.0	38.1
127	.35 Remington	1.0	5.0	14.4	32.3	62.5	2.5	11.0	27.2	51.5	82.3	1.1	5.2	13.8	29.3	54.1
128	.35 Remington	1.3	6.5	18.3	40.1	74.9	2.8	12.4	29.8	54.3	84.4	1.4	6.7	17.7	36.9	66.5
129	.350 Rem. Mag.	0.7	3.0	4.4	15.7	28.4	1.2	5.0	12.0	22.7	38.0	0.7	3.3	8.0	15.7	27.3
130	.350 Rem. Mag.	0.8	3.7	9.2	18.2	31.8	1.1	4.6	10.8	20.2	33.2	0.9	4.0	9.8	18.7	31.7
131	.351 Win. S.L.	1.5	7.2	19.6	41.2	74.8	2.6	11.1	26.2	47.3	73.2	1.7	7.6	19.5	39.3	69.1
132	.358 Winchester	0.8	3.7	9.8	20.4	37.8	1.5	6.5	15.7	30.0	50.0	0.9	4.0	10.0	19.9	35.3
133	.358 Winchester	1.0	4.4	11.3	22.9	40.9	1.4	5.9	14.2	28.6	43.7	1.1	4.8	11.8	23.0	39.6
134	.375 H&H Magnum	0.7	3.0	7.5	15.1	26.9	1.1	4.5	10.7	20.2	33.5	0.8	3.3	7.9	15.4	26.3
135	.375 H&H Magnum	0.8	3.4	8.6	17.4	31.0	1.2	5.0	11.8	22.3	37.0	0.9	3.7	9.0	17.6	30.2
136	.375 H&H Magnum	0.8	3.7	10.0	21.4	40.4	1.7	7.2	17.6	33.8	56.7	0.9	3.9	10.0	20.4	36.7
137	.38-40 Winchester	3.8	17.4	43.7	85.7	146.8	3.9	14.2	30.0	51.0	77.3	4.2	18.7	45.6	87.5	146.9
138	.38-55 Winchester	2.0	9.0	22.9	45.9	79.8	2.3	9.4	21.4	37.9	58.0	2.2	9.7	23.8	46.2	78.3
139	.44 Rem. Mag.	1.8	9.3	26.4	56.2	101.9	4.0	17.2	38.6	66.3	99.7	2.0	9.4	25.1	51.7	91.7
140	.44 Rem. Mag.	1.8	9.3	26.1	55.6	100.6	4.0	16.9	38.0	65.3	98.2	1.9	9.3	24.8	51.2	90.7
141	.44 Rem. Mag.	1.8	9.5	26.9	57.5	104.3	4.2	17.8	39.8	68.3	102.5	2.0	9.5	25.5	52.7	93.6
142	.44-40 Winchester	3.8	17.4	43.7	85.7	146.8	3.9	14.2	30.0	51.0	77.3	4.2	18.7	45.6	77.5	146.9
143	.444 Marlin	1.0	5.4	16.5	38.7	75.9	3.1	14.1	35.3	65.3	102.2	1.1	5.4	15.1	33.4	63.2
144	.45-70 Govt.	2.8	12.7	31.9	62.1	105.2	2.8	10.8	23.2	39.3	58.8	3.2	13.6	33.5	63.9	106.4
145	.458 Win. Mag.	1.2	5.3	13.4	26.9	47.7	1.5	6.3	15.0	27.9	45.3	1.3	5.7	14.0	27.2	46.5
146	.458 Win. Mag.	1.2	5.6	14.7	30.6	55.8	1.9	8.2	19.6	36.5	58.8	1.3	6.0	15.0	29.9	52.4

EXTERIOR BALLISTICS TABLES—RIFLE

Height of Trajectory (Inches) Above or Below Sight Line for Sights 0.9 Inches Above Bore Line

Line No.	Cartridge	Short Range (Yards)						Long Range (Yards)						
		50	100	150	200	250	300	100	150	200	250	300	400	500
121	.32-40 Winchester	0	-3.4	-11.8	-25.8	-46.2	-73.8	0	-6.7	-19.0	-37.7	-63.6	-139.6	-252.1
122	.338 Win. Mag.	0.5	0.9	0	-2.3	-6.1	-11.6	2.0	1.7	0	-3.2	-8.2	-24.3	-50.4
123	.338 Win. Mag.	0.2	0	-1.7	-5.2	-10.5	-18.0	2.6	2.1	0	-4.0	-10.2	-30.0	-61.9
124	.338 Win. Mag.	0.4	0	-2.3	-6.7	-13.5	-23.1	1.5	0	-3.6	-9.7	-18.6	-46.2	-90.7
125	.348 Winchester	0.3	0	-2.1	-6.2	-12.7	-21.9	1.4	0	-3.4	-9.2	-17.7	-44.4	-87.9
126	.348 Winchester	0.4	0	-2.2	-6.7	-13.8	-24.1	1.5	0	-3.7	-10.1	-19.6	-50.9	-103.8
127	.35 Remington	0.6	0	-3.0	-9.2	-19.1	-33.9	2.0	0	-5.1	-14.1	-27.8	-74.0	-152.3
128	.35 Remington	0.9	0	-4.1	-12.1	-25.1	-43.9	2.7	0	-6.7	-18.3	-35.8	-92.8	-185.5
129	.350 Rem. Mag.	0.2	0	-1.7	-5.1	-10.4	-17.9	2.6	2.1	0	-4.0	-10.3	-30.5	-64.0
130	.350 Rem. Mag.	0.4	0	-2.2	-6.4	-12.9	-21.9	1.5	0	-3.5	-9.2	-17.4	-42.5	-81.2
131	.351 Win. S.L.	0	-2.1	-7.8	-17.8	-32.9	-53.9	0	-4.7	-13.6	-27.6	-47.5	-108.8	-203.9
132	.358 Winchester	0.4	0	-2.2	-6.5	-13.3	-23.0	1.5	0	-3.6	-9.7	-18.6	-47.2	-94.1
133	.358 Winchester	0.5	0	-2.7	-7.9	-16.0	-27.1	1.8	0	-4.3	-11.4	-21.7	-53.5	-103.7
134	.375 H&H Magnum	0.2	0	-1.7	-5.1	-10.3	-17.6	2.5	2.1	0	-3.9	-10.0	-29.4	-60.7
135	.375 H&H Magnum	0.3	0	-2.0	-5.9	-11.9	-20.3	2.9	2.4	0	-4.5	-11.5	-33.8	-70.1
136	.375 H&H Magnum	0.3	0	-2.2	-6.5	-13.5	-23.4	1.5	0	-3.6	-9.8	-19.1	-49.1	-99.5
137	.38-40 Winchester	0	-6.7	-22.2	-47.3	-83.2	-130.8	0	-12.1	-33.9	-66.4	-110.6	-238.3	-425.6
138	.38-55 Winchester	0	-3.0	-10.5	-23.1	-41.3	-66.0	0	-6.0	-17.0	-33.8	-56.9	-125.0	-226.3
139	.44 Rem. Mag.	0	-2.7	-10.0	-23.2	-43.4	-71.9	0	-6.0	-17.8	-36.6	-63.6	-147.1	-276.2
140	.44 Rem. Mag.	0	-2.7	-10.0	-23.0	-43.0	-71.2	0	-5.9	-17.6	-36.3	-63.1	-145.4	-273.0
141	.44 Rem. Mag.	0	-2.7	-10.2	-23.6	-44.2	-73.3	0	-6.1	-18.1	-37.4	-65.1	-150.3	-282.5
142	.44-40 Winchester	0	-5.4	-18.5	-40.3	-71.9	-114.4	0	-10.3	-29.4	-58.3	-98.1	-214.7	-388.1
143	.444 Marlin	0.6	0	-3.2	-9.9	-21.3	-38.5	2.1	0	-5.6	-9.9	-15.9	-32.1	-87.8
144	.45-70 Govt.	0	-4.7	-15.8	-34.0	-60.0	-95.9	0	-8.7	-24.6	-48.2	-80.3	-172.4	-305.9
145	.458 Win. Mag.	0.7	0	-3.3	-9.6	-19.2	-32.5	2.2	0	-5.2	-13.6	-25.8	-63.2	-121.7
146	.458 Win. Mag.	0.8	0	-3.5	-10.3	-20.8	-35.6	2.4	0	-5.6	-14.9	-28.5	-71.5	-140.4

MAXIMUM RANGES—RIFLE

	Bullet		Bullet Type							
Cartridge	Wgt. Grs.	Ballistic Coefficient	I.V.I.	Federal	Rem-Peters	W-W	Muzzle Velocity (fps)	Maximum Range (yds.)	Velocity at Max. Range (fps)	Angle of Elevation (deg.)
6mm Remington	80	0.255			PSP,HPPL	PSP	3470	3731	371	32.9
6mm Remington	90	0.299			PSPCL		3190	4109	400	33.9
6mm Remington	100	0.356			PSPCL	PP(SP)	3130	4650	436	34.9
6.5mm M. Sch.	160	0.359	SP				2010	4119	423	36.1
6.5×55	160	0.359	SP				2390	4323	429	35.7
6.5mm Rem. Mag.	120	0.324			PSPCL		3210	3771	416	34.3
7mm Mauser	139	0.311	PSP				2660	4025	403	34.7
7mm Mauser	139	0.330		HSSP			2660	4197	415	35.0
7mm Mauser	160	0.274	KKSP				2520	3617	377	34.1
7mm Mauser	175	0.273	KK	HSSP	SP	SP	2440	3575	375	34.2
7mm Rem. Mag.	125	0.292		HSSP	PSPCL		3310	4079	396	33.7
7mm Rem. Mag.	150	0.346	SP	HSSP	PSPCL	PP(SP)	3110	4543	429	34.8
7mm Rem. Mag.	175	0.427		HSSP	PSPCL		2860	5166	474	36.1
7mm Rem. Mag.	175	0.273	SP			PP(SP)	2860	3729	379	33.8
8mm Mauser	170	0.205		HSSP	SPCL	PP(SP)	2360	2871	325	32.8
8mm Mauser*	170	0.312	PSP		HPPL		2510	3972	402	34.9
.17 Remington	25	0.151			HP		4040	2561	286	30.0
.218 Bee	46	0.130			SP,HP	OPE(HP)	2760	2095	261	30.2
.22 Hornet	45	0.130	PSP			SP	2690	2084	260	30.3
.22 Hornet	46	0.130				OPE	2690	2084	260	30.3
.22 Savage	70	0.249	PSP		HPPL		2760	3445	361	33.4
.222 Remington	50	0.188			PSP,MC		3140	2872	316	31.6
.222 Remington	50	0.175	PSP	SP			3140	2728	305	31.3
.222 Remington	55	0.272				PSP	3020	3774	380	33.6
.222 Rem. Mag.	55	0.201			HPPL	FMC	3240	3142	334	32.1
.222 Rem. Mag.	55	0.197	PSP		PSP		3240	3009	324	31.8
.223 Remington	55	0.272			PSP	FMC	3240	3843	381	33.4
.223 Remington	55	0.209		SP	HPPL		3240	3142	334	32.1
.223 Remington	55	0.197			PSP	PSP	3240	3067	324	31.8
.225 Winchester	55	0.208			PSP	PSP	3570	3212	335	31.8

Cartridge	Wgt. Grs.	Ballistic Coefficient	I.V.I.	Federal	Rem-Peters	W-W	Muzzle Velocity (fps)	Maximum Range (yds.)	Velocity at Max. Range (fps)	Angle of Elevation (deg.)
.22-250 Remington	55	0.230			HPPL	PSP	3730	3502	353	32.2
.22-250 Remington	55	0.197			PSP	PSP	3730	3112	327	31.4
.243 Winchester	75	0.236	PSP				3350	3478	356	32.6
.243 Winchester	80	0.255	PSP	SP	PSP	PSP	3350	3697	370	33.0
.243 Winchester	80	0.255		SP	HPPL		3350	3697	370	33.0
.243 Winchester	100	0.356	PSP	HSSP	PSPCL	PP(SP)	2960	4576	434	35.1
.25-06 Remington	87	0.263			HPPL	PSP	3440	3810	377	33.1
.25-06 Remington	87	0.230			HPPL		3440	3435	352	32.4
.25-06 Remington	90	0.259		HP	HPPL		3440	3768	374	33.0
.25-06 Remington	100	0.292			PSPCL	PEP	3230	4057	396	33.8

*Discontinued

MAXIMUM RANGES—RIFLE

Cartridge	Bullet		Bullet Type				Muzzle Velocity (fps)	Maximum Range (yds.)	Velocity at Max. Range (fps)	Angle of Elevation (deg.)
	Wgt. Grs.	Ballistic Coefficient	I.V.I.	Federal	Rem-Peters	W-W				
.25-06 Remington	117	0.349		HSSP		PEP	3060	4552	431	45.9
.25-06 Remington	120	0.362	PSP		PSPCL	PP(SP)	3010	4657	438	35.1
.25-06 Remington	120	0.362					3010	4657	438	35.1
.25-20 Winchester	86	0.190	SP		SP, Lead	SP, Lead	1460	2446	307	33.6
.25-35 Winchester	117	0.240			SPCL		2230	3183	351	33.8
.25-35 Winchester	117	0.213				SP	2230	2916	331	33.2
.250 Savage	87	0.263	SP			PSP	3030	3683	374	33.4
.250 Savage	100	0.285			PSP		2820	3836	387	34.0
.250 Savage	100	0.254	PSP			ST	2820	3516	365	33.4
.256 Win. Mag.	60	0.129				OPE(HP)	2760	2082	260	30.2
.257 Roberts	87	0.263				PSP	3170	3730	375	33.3
.257 Roberts	100	0.254			SPCL	ST	2900	3549	366	33.4
.257 Roberts	117	0.240			PSPCL	PP(SP)	2650	3324	354	33.3
.264 Win. Mag.	100	0.254			PSPCL	PSP	3320	3673	369	33.0
.264 Win. Mag.	140	0.385			PSPCL	PP(SP)	3030	4875	452	35.4
.270 Winchester	100	0.251		HSSP	PSP	PSP	3480	3687	368	32.8
.270 Winchester	130	0.372			OP	PP(SP)	3110	4793	445	35.2
.270 Winchester	130	0.336	PSP,ST		PSPCL	ST	3110	4454	423	34.6
.270 Winchester	150	0.345				P(SP)	2900	4442	426	35.0

MAXIMUM RANGES—RIFLE (continued)

Cartridge	Wgt. Grs.	Bullet Ballistic Coefficient	Bullet Type I.V.I.	Federal	Rem-Peters	W-W	Muzzle Velocity (fps)	Maximum Range (yds.)	Velocity at Max. Range (fps)	Angle of Elevation (deg.)
.270 Winchester	150	0.261		HSSP	SPCL		2900	3616	371	33.5
.270 Winchester	160	0.337	KKSP				2710	4283	419	35.1
.280 Remington	150	0.345			PSPCL		2890	4446	427	35.0
.280 Remington	165	0.290			SPCL		2820	3886	391	34.1
.284 Winchester	125	0.311				PP(SP)	3140	4216	407	34.2
.284 Winchester	150	0.344				PP(SP)	2860	4424	426	35.0
.30 Carbine	110	0.180				FMC	1990	2509	303	32.6
.30 Carbine	110	0.166		SP		HSP	1990	2369	291	32.2
.30 Remington	170	0.254			SPCL	ST	2120	3298	360	34.2
.30-30 Winchester	150	0.218	SP,ST,PNEU	HSSP		OPE,PP,ST	2390	3011	335	33.1
.30-30 Winchester	150	0.193			SPCL		2390	2750	316	32.5
.30-30 Winchester	170	0.254	KKSP,ST	HSSP	HPCL,SPCL	PP(SP),ST	2200	3307	360	34.1
.300 H&H Magnum	150	0.314				ST	3130	4242	409	34.3
.300 H&H Magnum	180	0.383			PSPCL	ST	2880	4790	449	35.5
.300 H&H Magnum	220	0.359				ST	2580	4426	431	35.5
.300 Win. Mag.	150	0.295		HSSP	PSPCL	PP(SP)	3290	4101	398	33.8
.300 Win. Mag.	180	0.438	ST	HSSP	PSPCL	PP(SP)	2960	5313	480	36.1
.300 Win. Mag.	220	0.380				ST	2680	4658	445	35.7
.300 Win. Mag.	220	0.294				PP(SP)	2680	3869	392	34.3
.30-06 Springfield	110	0.186				PSP	3380	2902	315	31.4
.30-06 Springfield	125	0.268		SP		PSP	3140	3710	378	33.4

MAXIMUM RANGES—RIFLE

Cartridge	Wgt. Grs.	Ballistic Coefficient	I.V.I.	Federal	Rem-Peters	W-W	Muzzle Velocity (fps)	Maximum Range (yds.)	Velocity at Max. Range (fps)	Angle of Elevation (deg.)
.30-06 Springfield	150	0.365			BP		2910	4635	439	35.3
.30-06 Springfield	150	0.365			MC		2810	4469	430	35.2
.30-06 Springfield	150	0.314	PSP,ST		PSPCL	ST	2910	4455	407	34.5
.30-06 Springfield	150	0.270				PP(SP)	2920	3721	378	33.7
.30-06 Springfield	165	0.468		BTSP			2800	5473	494	36.6
.30-06 Springfield	180	0.499				FMCBT	2700	5659	508	37.1
.30-06 Springfield	180	0.438				PP(SP)	2700	5162	477	36.4
.30-06 Springfield	180	0.412	OPE		BP		2700	4046	463	36.1
.30-06 Springfield	180	0.383	ST	HSSP	PSPCL	ST	2700	4697	447	35.7
.30-06 Springfield	180	0.248	KKSP		SPCL	PP(SP)	2700	3417	360	33.4
.30-06 Springfield	200	0.586		BTSP		ST	2550	6191	545	38.0
.30-06 Springfield	220	0.380			SPCL	PP(SP)	2410	4515	441	35.9
.30-06 Springfield	220	0.294	KKSP		PSPCL	ST	2430	3761	389	34.6
.30-40 Krag	180	0.383				PP(SP)	2430	4552	443	36.0
.30-40 Krag	180	0.248			SPCL		2430	3327	358	33.7
.30-40 Krag	220	0.382	PSP,ST		PSPCL	ST	2160	4386	438	36.3
.300 Savage	150	0.314		HSSP	SPCL	ST	2630	4039	404	34.7
.300 Savage	150	0.270			PSPCL	PP(SP)	2630	3620	375	34.0
.300 Savage	150	0.223	ST				2630	3141	341	33.0
.300 Savage	180	0.383	KKSP	HSSP		ST	2350	4506	442	36.1
.300 Savage	180	0.248			SPCL		2350	3299	357	33.8
.300 Savage	180	0.246	KKSP		SPCL	PP(SP)	2120	3204	354	34.1
.303 Savage	190	0.252	PSP,ST			ST	1940	3182	356	34.4
.303 British	150	0.300	OPE				2700	3930	396	34.4
.303 British	180	0.407		HSSP			2460	4715	457	36.3
.303 British	180	0.369	ST		SPCL		2460	4450	436	35.7
.303 British	180	0.246	KKSP				2460	3324	357	33.7
.303 British	180	0.237	KKSP			PP(SP)	2460	3229	350	33.5
.303 British	215	0.285			SP		2170	3582	381	34.8
.308 Winchester	110	0.186				PSP	3180	2861	314	31.5
.308 Winchester	125	0.268				PSP	3100	3754	377	33.5

| | | Bullet | | | | Bullet Type | | | | |
Cartridge	Wgt. Grs.	Ballistic Coefficient	I.V.I.	Federal	Rem-Peters	W-W	Muzzle Velocity (fps)	Maximum Range (yds.)	Velocity at Max. Range (fps)	Angle of Elevation (deg.)
.308 Winchester	150	0.352	PSP,ST	HSSP	MC	ST	2810	4470	430	35.2
.308 Winchester	150	0.314			PSPCL	PP(SP)	2820	4419	406	34.5
.308 Winchester	150	0.270				ST	2820	3587	377	33.8
.308 Winchester	180	0.303	ST	HSSP	PSPCL		2620	4655	446	35.8
.308 Winchester	180	0.248			SPCL	PP(SP)	2620	3392	359	33.5
.308 Winchester	180	0.237	KKSP			ST	2620	3284	352	33.5
.308 Winchester	200	0.345	KKSP			ST	2450	4235	421	35.4
.32 Remington	170	0.213					2140	2883	330	33.3
.32 Win. Spl.	170	0.239	KKSP,ST	HSSP	HPCL,SPCL		2250	3176	350	33.7

MAXIMUM RANGES—RIFLE

| | | Bullet | | | | Bullet Type | | | | |
Cartridge	Wgt. Grs.	Ballistic Coefficient	I.V.I.	Federal	Rem-Peters	W-W	Muzzle Velocity (fps)	Maximum Range (yds.)	Velocity at Max. Range (fps)	Angle of Elevation (deg.)
.32 Win. Special	170	0.205				PP(SP),ST	2250	2840	325	33.0
.32-20 Winchester	100	0.166			SP, Lead	SP, Lead	1210	2122	284	33.4
.32-20 Winchester	115	0.200	SP				1490	2540	314	33.8
.32-40 Winchester	170	0.279	KKSP				1530	3227	369	35.5
.338 Win. Mag.	200	0.308				PP(SP)	2960	4115	404	34.3
.338 Win. Mag.	250	0.329				ST,PP(SP)	2660	4192	414	35.0
.338 Win. Mag.	300	0.297				PP(SP)	2430	3798	392	34.7
.348 Winchester	200	0.276				ST	2520	3633	378	34.2
.35 Remington	150	0.184		HSSP	PSPCL		2300	2633	308	32.4
.35 Remington	200	0.193	SP		SPCL	PP(SP),ST	2020	2648	313	32.9
.350 Rem. Mag.	200	0.294			PSPCL		2710	3878	392	34.3

Cartridge										
.351 Win. S.L.	180	0.233			SP	1850	2977	342	34.1
.358 Winchester	200	0.262			ST	2490	3482	368	33.9
.358 Winchester	250	0.326			ST	2230	3994	407	35.4
.375 H&H Magnum	270	0.326	SP		PP(SP)	2690	4172	412	34.9
.375 H&H Magnum	300	0.324	MC		ST	2530	4088	410	35.0
.375 H&H Magnum	300	0.234	SP		FMC	2530	3219	348	33.3
.38-40 Winchester	180	0.171			SP	1160	2147	288	33.7
.38-55 Winchester	255	0.311	SP			1590	3505	389	35.9
.44 Remington	240	0.163	SP			1760	2274	287	32.4
.44 Rem. Mag.	240	0.166	SP			1760	2305	289	32.5
.44 Rem. Mag.	240	0.158		HSP	HSP	1760	2217	282	32.3
.44-40 Winchester	200	0.160	SP		SP	1190	2063	279	33.3
.444 Marlin	240	0.146	SP			2350	2222	275	31.2
.45-70 Govt.	405	0.280			SP	1330	3119	366	35.8
.458 Win. Mag.	500	0.345	MC		FMC	2040	4026	416	35.9
.458 Win. Mag.	510	0.274	SP		SP	2040	3427	372	34.8

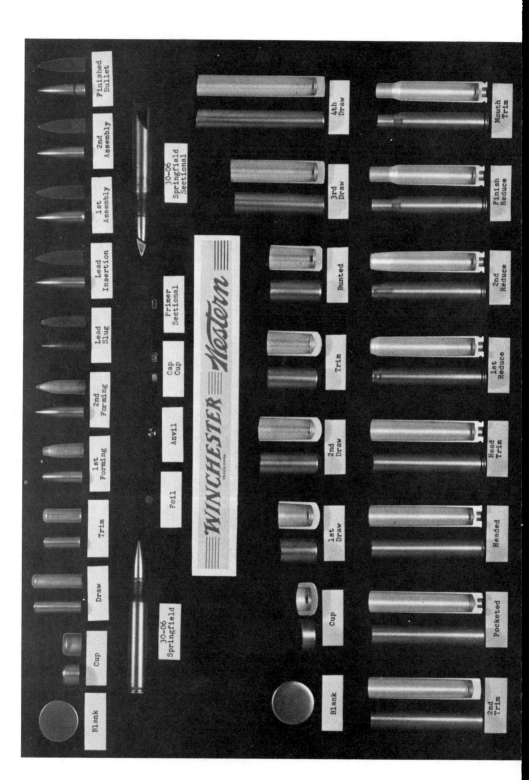

Blank · Cup · Draw · Trim · 1st Forming · 2nd Forming · Lead Slug · Lead Insertion · 1st Assembly · 2nd Assembly · Finished Bullet

30-06 Springfield

Foil · Anvil · Cap Cup · Primer Sectional

WINCHESTER Western
TRADE-MARK

30-06 Springfield Sectional

Blank · Cup · 1st Draw · 2nd Draw · Trim · Bunted · 3rd Draw · 4th Draw

2nd Trim · Pocketed · Headed · Head Trim · 1st Reduce · 2nd Reduce · Finish Reduce · Mouth Trim

CHAPTER SIX

Choosing a Centerfire Handgun Cartridge

Handgun cartridges, for the most part, have evolved as cartridges to be used for personal defense or anti-personnel military rounds and, of course, as police cartridges. Relatively few have been developed with only the hunter/sportsman in mind.

Among those developed for personal protection there are a fair number of cartridges that are totally useless for the intended purpose (or for that matter any other purpose).

Once these useless cartridges have been eliminated, the selection of handgun calibers becomes comparatively simple. In the interest of clarity, we will first discuss those calibers which have no real application except perhaps for plinking at tin cans. With the price of centerfire ammunition, it would be far more practical to choose a .22 rimfire for such an application.

CENTERFIRE HANDGUN CARTRIDGES FOR WHICH THERE ARE NO PRACTICAL APPLICATIONS

.25 CALIBER

.25 Auto
50-Grain Bullet
The .25 Auto was created with the maximum in concealment for a handgun as the criterion. While most good .25 Autos are readily concealable, the ballistics of the cartridge are so low that one runs a grave risk of merely antagonizing an opponent should the gun be used for personal protection.

The relatively poor accuracy and the full metal case bullet make the cartridge useless to the hunter/sportsman.

.30 CALIBER

.30 Luger
93-Grain Bullet
Originally designed as a military cartridge, the .30 Luger was a good one in its day. However, for modern military purposes, the 9mm Luger will outperform the .30 Luger.

The full metal case bullet makes it totally unsuitable for sporting or police purposes.

.32 CALIBER

.32 Short Colt
80-Grain Bullet
The ballistics of this round make it useful only for the generation of low-level noise.

.32 Long Colt
82-Grain Bullet
The 2-grain heavier bullet than the short version and the 10 fps increase in velocity give this cartridge absolutely no advantage over the .32 Short Colt. It is another useless round at best.

.32 Colt New Police
98-Grain Bullet
The increase in bullet weight (over the .32 Short or Long Colt cartridges) is offset by a decrease in velocity. The result is that the .32 C.N.P. is just as useless as its ballistic twins.

.32 S&W
88-Grain Bullet
This cartridge fits into the same ballistic category as the previous .32's and is every bit as inefficient for any application.

.32 S&W Long
98-Grain Bullet
While this round offers approximately a 15% increase in muzzle energy over the previous .32's it is still a ballistic midget and unsuitable for any application.

.32 Auto
71-Grain Bullet
The .32 Auto is more potent than any of the aforementioned .32's but still too light for any purpose. Some foreign target handguns have been chambered for this round. As a paper puncher in a high-grade target pistol, it may serve well, but the lack of match-grade ammo handicaps the round for such an application.

.38 CALIBER

.38 Short Colt
125-Grain Bullet
Another ballistic midget with no practical application.

.38 Long Colt
150-Grain Bullet
Still we are short of enough punch to make this round useful.

.38 Colt New Police
150-Grain Bullet
With respect to energy, this round is practically a ballistic twin to the .38 Long Colt. Hence, no justifiable applications exist for this cartridge.

.38 S&W
145- and 146-Grain Bullets
This cartridge uses a case that is identical to the .38 Colt New Police. As can be expected, ballistics are also identical. As a result, the cartridge cannot be recommended for any purpose despite its use by some law-enforcement agencies.

All of the remaining handgun cartridges have valid applications. Handgun cartridge selection in centerfire calibers should, therefore, be limited to the following 18 cartridges. Calibers will be shown in ascending numerical order with the available bullet weights matched to their applications. It is hoped that the comments listed will help the reader in making his choice(s).

CENTERFIRE HANDGUN CARTRIDGE SELECTOR GUIDE

.22 CALIBER

.22 Remington Jet
40-Grain Bullet—Varmint
This varmint round has never proved too popular despite its reasonably good accuracy. Extraction problems from the cylinder are undoubtedly the cause of its low popularity. The chambers of the revolver's cylinder must be kept completely free from any oil or solvent as must the cases. Even when great care is taken, extraction is, at best, difficult. While the cartridge will perform well on varmint up to 80 yards, such difficulty really cannot make the gun a desirable selection.

.221 Remington Fireball
50-Grain Bullet—Varmint
This very accurate round is available only in a single-shot, bolt-action pistol. It performs excellently at ranges up to 100 yards on all sizes of varmints.

.25 CALIBER

.256 Winchester Magnum
60-Grain Bullet—Varmint
This is perhaps one of the best varmint handgun cartridges ever to come down the pike. However, the only handgun ever chambered for it was the Ruger Hawkeye.

The Hawkeye was a single-shot pistol capable of a very high degree of accuracy with the .256 cartridge. However, it never proved very popular, and the gun was quickly discontinued. I believe it simply was about five years ahead of the shooting public.

If you are lucky enough to find a used Hawkeye it makes an excellent varmint handgun to about 100 yards. Perhaps it makes the very best selection. But most Hawkeyes are bought up by collectors at very high prices. The .256 is an extremely difficult cartridge to reload. In fact, it is almost impossible to handload successfully for the .256 Winchester Magnum.

.32 CALIBER

.32-20 Winchester
100-Grain Bullet—Small Game, Varmint, Self-Protection

The .32-20 is a forgotten round, and no handguns have been chambered for it in many years.

However, the ballistics are such that it can be used for small game and varmint to about 80 yards. It also has enough energy to make it a practical round for self-defense.

.38 CALIBER

9mm Luger
**95-Grain Bullet | Varmint, Small Game,
100-Grain Bullet | Self-Protection, Police**

**115-Grain Bullet |
124-Grain Bullet | Military**

The 95- and 100-grain (soft-point or hollow-point) bullets make excellent rounds for all handgun applications. The cartridge offers excellent ballistic levels combined with modest recoil and noise. For those who favor semiautomatic handguns, this is the number-one choice.

The full metal case bullets (115 and 124 grain) have no application other than for military purposes or perhaps target shooting. However, the lack of match-grade ammunition has kept the 9mm Luger from any real application as a target-shooting handgun.

.357 Magnum
**110-Grain Bullet | Varmint, Self-Protection,
125-Grain Bullet | Small Game, Police
158-Grain Bullet |**

The .357 Magnum handgun is perhaps the most versatile of all handgun calibers. While it can be used for all the listed purposes, its ability to use .38 Special ammunition also makes it an excellent target-shooting gun.

Some shooters have successfully used the 158-grain jacketed hollow points on light big game. However, at the risk of great outcries from serious handgun hunters, big-game shooting with a handgun falls into the category of stunt shooting. One would be better armed with almost any rifle, for such purposes.

The best small-game load for the .357 Magnum revolver continues to be a .38 Special 148-grain wadcutter round.

It is popular with the 110-grain bullet for personal protection or police work.

The muzzle blast and recoil are somewhat heavy, but most experienced handgunners can handle the .357 Magnum very well. It is perhaps the *heaviest* handgun load most shooters can master.

Due to its ability to handle all .38 Special loads, the .357 Magnum revolver is perhaps the one all-purpose handgun. A shooter can use .38 Special loads for 90% of his shooting, but when heavier loads are needed the .357 will gracefully accommodate your needs. The .357 revolver should be the number-one choice of handguns for shooters who require more than a straight target gun or a basic self-protection or police firearm. It is too heavy for police work where very extensive training is not available.

.38 Special

148-Grain Bullet { Target, Small Game, Self-Protection

158-Grain Bullet { Police, Military, Self-Protection

200-Grain Bullet—N.R.

95-Grain +P Bullet } Varmint,
110-Grain +P Bullet { Small Game,
125-Grain +P Bullet { Self-Protection,
150-Grain +P Bullet } Police

158-Grain +P Bullet—Police, Self-Protection

Unquestionably the most popular handgun cartridge, the .38 Special has been both praised and damned. However, the damns, for the most part, come from those not in a position to be qualified to make a total judgment. Most of the bad mouthing concerns the use of the .38 Special with a 158-grain round-nose bullet for police work.

The 158-grain round-nose bullet and the 200-grain round-nose bullet are of little use.

The 158-grain semi-wadcutter lead and hollow-point lead bullets are, however, very useful police rounds. The +P rounds offer additional advantages where the police officer receives extensive and repetitive training. With the proper ammunition, the .38 Special should remain the number-one selection for police work. It offers an ideal compromise of moderate noise level, low muzzle flashes, moderate recoil, excellent accuracy, and easy mastering by the average police officer who receives sufficient and repetitive training. All this is combined with enough punch to get the job done (with the right load) without overly jeopardizing innocent bystanders.

The 148-grain match-grade wadcutter is still the best target load available to the serious shooter. It also makes a good small-game load, and for the one-load shooter, it will make a good self-defense load. It will certainly outperform the 158- or 200-grain round-nose bullets with respect to stopping a hostile adversary.

The Plus P loads (+P) add a big area of versatility to the .38 Special. However, these loads are loaded to higher than normal pressures and should be used only in guns that the manufacturer states are suitable for use with such ammunition.

The 95-grain +P bullet is designed for ideal performances in 2-inch barrels.

The 110-grain +P bullet is the most excellent self-defense, police, and varmint load available in the .38 Special. It will also work on small game but may damage more meat than you might consider ideal. As a police load, the 110-grain +P round requires a bit more training than standard loads. Noise and muzzle flash are noticeably increased. This load seems to perform best in 4-inch barrels.

The 125-grain is another good choice for a police, varmint, and self-protection load. The 150-grain and 158-grain semi-wadcutters and lead hollow points are also excellent choices for police work. The 158-grain and 200-grain round-nose bullets are poor on performance level.

For those who do not have the occasional need for the extra power and range of the .357 Magnum, the .38 Special is the very best choice that can be made for a handgun caliber. The only exception would be for the few handgunners who really can handle large calibers very well and who have a need for maximum stopping power. To even start to qualify for this group one must shoot at least 50 rounds of ammunition each and every week with the chosen caliber and gun.

.38 Auto
130-Grain Bullet—N.R.
Although this cartridge possesses excellent paper ballistics, it is sorely lacking in accuracy. Additionally, it is a dead cartridge since no guns are manufactured for it. The exception is, of course, that the .38 Auto will fire in the .38 Super Chamber. Actually, it has the same case size but is loaded to a lower ballistic level than the Super. The 130-grain metal case bullet makes it suitable for military purposes only.

.38 Super Auto
125-Grain Bullet—N.R.
130-Grain Bullet—N.R.
The paper ballistics are excellent for this cartridge, but rare indeed is a .38 Super that will shoot accurately. With the 125-grain hollow-point bullet, it would be an excellent varmint, small-game, and self-protection load if reasonable accuracy was obtainable.

The cartridge/gun combination is also plagued with a pierced-primer problem.

The 130-grain bullet would be suitable only to military applications, because it is available only in a full metal case configuration.

.380 Auto
95-Grain Bullet—Self-Protection
This is the very lightest cartridge that should be used for self-protection. The soft-point bullets will not offer much advantage over

the full metal case bullets. This is due to the fact that the relatively low velocity of the cartridge will not allow good or reliable expansion of a soft-point bullet.

This cartridge is borderline, and one should weigh this factor when considering it for self-defense.

Its light recoil and low noise level allow for easy mastering. If an automatic is what you want and you also desire minimum recoil combined with easy concealment, this caliber might be just what you are looking for. Beware, though, of unreliable firearms.

.38-40 Winchester
180-Grain Bullet—Self-Defense
This obsolete round offers more than sufficient ballistics for self-defense. However, it's not an easy cartridge to master. It is actually a .40-caliber cartridge, despite the misleading nomenclature.

No guns are currently chambered for it, and ammo is not easy to find. It is a dead cartridge, and there is no reason to wish it to be anything else.

.41 CALIBER

.41 Remington Magnum
210-Grain Bullet—Varmint, Self-Defense
The 210-grain jacketed bullet makes an excellent varmint load. However, it does not seem to perform as well as some .357 Magnum loads. Perhaps this problem lies in the shooter's ability to handle the gun. Some have used this same load successfully on light big game. More than likely, more have used it unsuccessfully. Almost any rifle cartridge would give the hunter a far better edge and the game a more sporting chance at a swift death or a non-maimed life. Recoil and muzzle blast are both quite heavy.

The 210-grain load with the lead bullet was supposed to be an intermediate police load. I guess the designers simply didn't understand the police market. It is simply too much cartridge for 97% of the peace officers in the country.

.44 CALIBER

.44 S&W Special
246-Grain Bullet—Self-Protection
The .44 Special is an excellent performer on paper; however, the only current factory load is a round-nose bullet. This bullet configuration is poor. The results are that actual performance is well below that which the published ballistics would lead one to expect. Handloading wadcutters, semi-wadcutters, or hollow points will increase its potential substantially.

It does, however, have enough to get the job done if the shooter can handle the recoil.

.44 Remington Magnum
240-Grain Bullet

No applications are listed for this cartridge. It is certainly capable of doing anything any handgun can, because it is, without a doubt, the most powerful factory cartridge available for a handgun.

The point here, however, is that it is simply too powerful. Very few shooters can master it. The .44 Magnum has turned more handgun shooters into flinchers than any other caliber. The recoil and muzzle blast are far in excess of being uncomfortable. Painful would perhaps describe it better. Ninety percent of all .44 Magnum shooters turn to reloading with reduced loads in order to shoot their handgun.

The .44 Magnum's great popularity may be the result of a "Cadillac" syndrome. The purchaser simply may want to own the "world's most powerful handgun." Or it may be some sort of macho trip.

Or it could be the result of the big "IF." IF you need the most powerful handgun in the world. IF you have bears chasing you home and IF you can't carry a rifle. IF you are not bothered by excessive recoil or IF you can stand painful noise. IF you can manage to shoot well with the deck stacked against you. Or IF you feel it makes sense to buy a powerful handgun and shoot only reloaded ammunition that runs the cartridge at one half to two-thirds throttle. If any of the above applies, then the .44 Magnum could be a good selection.

This condemnation is given primarily to make the average shooter aware that the .44 Magnum is not actually a very useful cartridge. For the chosen few, and they do exist, the .44 Magnum can do it all.

The handgunner should keep in mind that the 240-grain lead bullet develops 230 foot-pounds more muzzle energy than the 240-grain jacketed bullets in handgun length barrels. Many .44 Magnum shooters are greatly impressed by the appearance of the various jacketed bullet loads, never realizing that they leave the muzzle of a handgun at about 170 fps less than their lead counterparts. However, the jacketed bullets do not leave as hard a gun-cleaning chore when you are through shooting.

.44-40 Winchester
200-Grain Bullet—Self-Protection

This obsolete and dead cartridge offers more than sufficient ballistics for self-protection. Recoil is heavy, too heavy for most shooters. The difficulty in finding ammunition does not really make it a good selection. The jacketed soft points used in factory ammunition will not expand at handgun velocity.

.45 CALIBER

.45 Auto
185-Grain Bullet—Target, Self-Protection,
Police, Varmint, Small Game
230-Grain Bullet—Self-Protection, Police, Military

Recoil of the .45 Auto cartridge in a semiautomatic pistol is not as bad as many have preached. It is heavy, but it can be mastered by an experienced shooter without great difficulty. It has less recoil than most .357 Magnum loads.

The 185-grain wadcutter bullet is a great favorite for target shooting and small-game shooting. This load, however, seldom feeds well except in guns that have been carefully altered for target shooting.

The 185-grain hollow point serves well for self-defense, police (where the officers are well trained), and varmint shooting. The .45 Auto is not an ideal police firearm and should not be considered where the peace officer is already using a better round for his purpose, such as the .38 Special +P, .357 Magnum, or 9mm Luger.

The 230-grain full metal case bullet makes itself useful as a self-protection, police, or military round.

.45 Auto Rim
230-Grain Bullet—Self-Protection
The round-nose lead bullet used in this round affords ample ballistics for self-protection. This cartridge is a rimmed version of the .45 Auto designed for use in revolvers. The .45 Auto loads can also be used in revolvers. This requires the use of special half-moon clips. The Auto Rim cartridge serves no purpose that cannot be better filled by another round.

.45 Colt
255-Grain Bullet—Self-Protection
While this cartridge affords the shooter sufficient paper ballistics for self-protection, its poor nose profile cuts its actual performance below anticipated levels. Reloaded with semi-wadcutter bullets, it is about the maximum cartridge the average shooter can handle with respect to recoil.

The author has taken a very great number of small-game animals and varmints with this cartridge, but there are a great many cartridges better suited for those purposes.

GENERAL HANDGUN OPINION
It should be kept in mind that it is no secret that the author has had a strong personal love affair with the following six good automatic pistols.

> The S&W 39 in 9mm Luger
> The S&W 59 in 9mm Luger
> The Colt 1911 in .45 ACP
> The Colt Commander in .45 ACP
> The Walther PP in .380 Auto
> The Walther PPK in .380 Auto

This affair is over!

The author has carried the above guns for almost every conceivable purpose. *However*, semiautomatic pistols, despite having been number one with the author, are far from the ideal police handguns. They are slower to bring into play, subject to ammunition malfunctions, and require a great deal more training time than even the biggest departments can afford their officers. Automatic pistols are acceptable for police work only for individuals who qualify as gun experts, who fire a minimum of 50 rounds per week, and who clean their firearm thoroughly after each and every use.

Additionally, a highly skilled armorer should inspect and repair all automatics used as police firearms on a very frequent basis.

While the author has carried automatics, he *strongly* suggests this is not a good course for any police officer who cannot fit the above criteria. The author has not always felt this way. He initially thought a revolver was best for police work, went through what he believed was an educated change to good automatics, and once again came to the final realization that revolvers are the best general police firearm. The person most responsible for this final change was Lt. Frank McGee, Commanding Officer of Firearms and Tactics Section for the New York City Police Department.

The N.Y.P.D. has at their disposal, the most extensive files and research on shootings involving police officers that exist today. These case files on over 5000 shooting incidents involving police officers prove beyond a doubt that the revolver is the only way to travel for the average police officer.

Those who have the responsibility for arming police departments may find it beneficial to contact the N.Y.P.D. (on their official stationary) through Lt. McGee at the Firearms and Tactics Section, Police Academy, City Island, Bronx, N.Y. 10461. A trip to New York City to discuss the matter would be the most practical step one could make before arming, or rearming, a police department. Lt. Frank McGee is the expert!

CHAPTER SEVEN

Centerfire Exterior Ballistics Tables for Handguns

The centerfire ballistics tables for handguns reflect, wherever practical, velocities obtained in a 4-inch vented test barrel. This new system, developed by Remington Firearms Company, simulates the actual revolver system with a cylinder-to-barrel gap. Ballistics obtained in such barrels very nearly duplicate the results that the shooter can expect to obtain in his revolver.

It is unfortunate that data for all revolver cartridges using this system is not yet available. However, most of the popular revolver calibers have been tested under this new system.

Additionally, a long-range ballistics table has been included for popular handgun cartridges for ranges up to 200 yards. This data will be of particular interest to silhouette shooters. The data for the long-range tables was developed by the ammunition laboratory at Winchester-Western. Maximum-range distance is listed for as many loads as data has been developed.

As in both the rimfire and centerfire rifle ballistics tables, the numbers in the handgun tables must not be considered as absolutes. The numbers are average numbers based on the results one might expect from the average of all the possible variables.

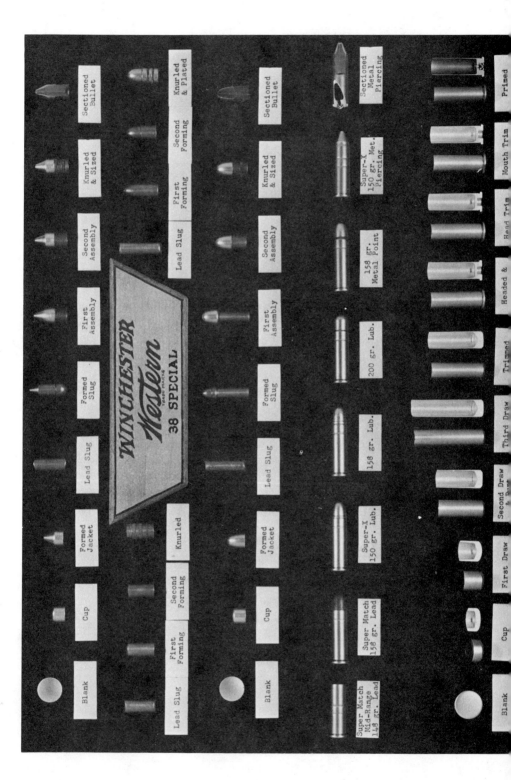

EXTERIOR BALLISTICS TABLES—HANDGUN

Cartridge	Bullet Wgt. Grs.	Type	Velocity in fps 0 Yds.	50 Yds.	100 Yds.	Energy in ft-lbs. 0 Yds.	50 Yds.	100 Yds.	Midrange Traj. in Inches From Bore Line 50-yd. Range	100-yd. Range	Pistol or Revolver Bbl. Length Inches*
9mm Luger	90	All	1330	1070	935	353	229	175	0.8	3.6	4
9mm Luger	100	S.P.	1320	1114	991	387	275	218	0.7	3.4	4
9mm Luger	115	All	1155	1047	971	341	280	241	0.9	3.9	4
9mm Luger	124	All	1110	1030	971	339	292	259	1.0	4.1	4
.22 Rem. Jet CF Mag.	40	All	2100	1790	1510	390	285	200	0.3	1.4	8⅜
.221 Rem. Fireball	50	All	2650	2380	2130	780	630	505	0.2	0.8	10½
.25 Auto	50	All	810	755	700	73	63	54	1.8	7.7	2
.256 Win. Mag.	60	All	2350	2030	1760	735	550	415	0.3	1.1	8½
.30 Luger	93	All	1220	1110	1040	305	255	225	0.9	3.5	4½
.32 Short Colt	80	All	745	665	590	100	79	62	2.2	9.9	4
.32 Long Colt	82	All	755	715	675	100	93	83	2.0	8.7	4
.32 Colt N.P.	98	All	680	635	595	100	88	77	2.5	11.0	4
.32 S&W	88	All	680	645	610	90	81	73	2.5	10.5	3
.32 S&W Long	98	All	705	670	635	115	98	88	2.3	10.5	4
.32 Auto	71	All	905	855	810	129	115	97	1.4	5.8	4
.32-20 Winchester	100	All	1030	970	920	270	210	190	1.2	4.4	6
.357 Magnum	110	All	1295	1094	975	410	292	232	0.8	3.5	4V
.357 Magnum	125	All	1450	1240	1090	583	427	330	0.6	2.8	4V
.357 Magnum	158	All	1235	1104	1015	535	428	361	0.8	3.5	4V
.38 Short Colt	125	All	730	685	645	150	130	115	2.2	9.4	6
.38 Long Colt	150	All	730	700	670	175	165	150	2.1	8.8	6

*"V" indicates vented barrel (simulates revolver system with cylinder-to-barrel gap).

EXTERIOR BALLISTICS TABLES—HANDGUN

Cartridge	Bullet Wgt. Grs.	Bullet Type	Velocity in fps 0 Yds.	50 Yds.	100 Yds.	Energy in ft.-lbs. 0 Yds.	50 Yds.	100 Yds.	Midrange Traj. in Inches From Bore Line 50-yd. Range	100-yd. Range	Pistol or Revolver Bbl. Length Inches*
.38 Colt N.P.	150	All	680	645	615	155	140	125	2.5	10.5	4
.38 S&W	145-146	All	685	650	620	150	135	125	2.4	10.0	4
.38 Special +P	95	All	1175	1044	959	291	230	194	0.9	3.9	4V
.38 Special +P	110	All	1020	945	887	254	218	192	1.1	4.8	4V
.38 Special +P	125	All	945	898	858	248	224	204	1.3	5.4	4V
.38 Special +P	150	All	910	870	835	276	252	232	1.4	5.7	4V
.38 Special	158	All	755	723	693	200	183	168	2.0	8.3	4V
.38 Special +P	158	All	915	878	844	294	270	250	1.4	5.6	4V
.38 Special	200	All	630	614	594	179	168	157	2.8	11.5	4V
.38 Auto	130	All	1040	980	925	310	275*	245	1.0	4.7	4½
.38 Super Auto +P	130	All	1245	1120	1035	450	365	310	0.8	3.4	5
.380 Auto	95	MC	955	865	785	190	160	130	1.4	5.9	3¾
.38-40 Winchester	180	SP	975	920	870	380	340	300	1.5	5.4	5
.41 Magnum	210	SP	1300	1162	1062	778	630	526	0.7	3.2	4V
.41 Magnum	210	Lead	965	898	842	434	376	331	1.3	5.4	4V
.44 S&W Spl.	246	All	755	725	695	310	285	265	2.0	8.3	6½
.44 Rem. Mag.	240	Jack.	1180	1081	1010	741	623	543	0.9	3.7	4V
.44 Rem. Mag.	240	Lead	1350	1200	1085	970	765	630	0.7	3.0	4V
.44-40 Winchester	200	SP	975	920	865	420	375	330	0.5	5.7	7½
.45 Colt	255	All	860	820	780	420	380	345	1.5	6.5	5½
.45 Auto	230	All	810	776	745	335	308	284	1.7	7.2	5
.45 Auto	185	HP	940	890	846	363	325	294	1.3	5.5	5

*"V" indicates vented barrel (simulates revolver system with cylinder-to-barrel gap).

EXTERIOR BALLISTICS TABLES—HANDGUN

Cartridge	Bullet		Velocity in fps			Energy in ft.-lbs.			Midrange Traj. in Inches From Bore Line		Pistol or Revolver
	Wgt. Grs.	Type	0 Yds.	50 Yds.	100 Yds.	0 Yds.	50 Yds.	100 Yds.	50-yd. Range	100-yd. Range	Bbl. Length Inches*
.45 Auto	230	M. Pen.	945	890	835	455	405	355	1.3	5.4	5
.45 Auto Rim	230	Lead	810	770	730	335	305	270	1.8	7.4	5½
Match											
.38 Special Match	148	All	710	634	566	166	132	105	2.4	10.8	4V
.45 Auto Match	185	All	770	707	650	244	205	174	2.0	8.7	5

*"V" indicates vented barrel (simulates revolver system with cylinder-to-barrel gap).

EXTERIOR BALLISTICS TABLES—HANDGUN
APPROXIMATE MAXIMUM HORIZONTAL DISTANCE TO POINT OF FIRST IMPACT*

Cartridge	Bullet Wgt. Grs.	Distance in Yds.	Cartridge	Bullet Wgt. Grs.	Distance in Yds.
9mm Luger	115	1867	.38 Special	200	1934
9mm Luger	124	1900	.38 Super Auto +P	130	2034
.22 Rem. Jet CF Mag.	40	2334	.380 Auto	95	1467
.221 Rem. Fireball	50	2667			
.25 Auto	50	1400	.38-40 Winchester	180	2034
.256 Win. Mag.	60	2167	.41 Magnum (Soft Point)	210	2367
.30 Luger	93	1900	.41 Magnum (Lead)	210	2234
			.44 S&W Special	246	1734
.32 Short Colt	80	1400			
			.44 Rem. Mag.	240	2500
.32 Long Colt	82	1400	.44-40 Winchester	200	2000
.32 Colt N.P.	100-98	1400	.45 Colt	250	1800
.32 S&W	88	1334	.45 Auto	230	1700
.32 S&W Long	98	1434			
.32 Auto	71	1467	.45 Auto Rim	230	1634
.32-20 Winchester	100	1900			

EXTERIOR BALLISTICS TABLES—HANDGUN (continued)
APPROXIMATE MAXIMUM HORIZONTAL DISTANCE TO POINT OF FIRST IMPACT*

Cartridge	Bullet Wgt. Grs.	Distance in Yds.	Cartridge	Bullet Wgt. Grs.	Distance in Yds.
.38 Long Colt	150	1534			
.38 Colt N.P.	150	1434	*Match*		
.38 S&W	146	1467			
.38 Special +P	110	1800	.38 Special Match	148	1634
.38 Special +P	150	2100	.38 Special Match	158	1834
.38 Special	158	1834	.45 Auto Match	185	1467

*Calculated for bullets with tangent ogive nose of .8-caliber radius, cylindrical bearing, and flat base with nominal velocities. The optimum angle of departure for maximum horizontal range is approximately 35 degrees.

LONG-RANGE EXTERIOR BALLISTICS TABLES—HANDGUN

Caliber	Bullet	Velocity (fps)					Bullet Drop (Inches)				Barrel Length (Inches)*
		Muzzle	25 Yds.	50 Yds.	100 Yds.	200 Yds.	25 Yds.	50 Yds.	100 Yds.	200 Yds.	
.357 Magnum	158 JSP	1235	1165	1104	1015	900	0.7	3.1	13.0	59.0	4V
9mm Luger	115 FMC	1155	1095	1047	970	865	0.8	3.5	15.0	65.5	4
.38 Special +P	158 SWC	915	895	878	845	785	1.3	5.3	22.0	92.5	4V
.38 Super Auto +P	130 FMC	1245	1180	1120	1035	920	0.7	3.0	13.0	57.0	5
.41 Rem. Mag.	210 JSP	1300	1225	1162	1060	935	0.7	2.8	12.0	53.5	4V
.44 Rem. Mag.	240 HSP	1350	1270	1200	1085	950	0.6	2.6	11.0	50.5	4V
.45 Colt	255 Lead	860	840	820	780	710	1.5	6.1	25.0	107.0	5½
.45 Auto	230 FMC	810	795	776	745	690	1.7	6.8	28.0	119.0	5

Caliber	Bullet	Vel. at Muzzle	Energy (ft.-lbs.)					Midrange Trajectory (Inches)				Barrel Length (Inches)*
			Muzzle	25 Yds.	50 Yds.	100 Yds.	200 Yds.	25 Yds.	50 Yds.	100 Yds.	200 Yds.	
.357 Magnum	158 JSP	1235	535	475	430	360	280	0.2	0.8	3.5	16.0	4V
9mm Luger	115 FMC	1155	340	305	280	240	190	0.2	0.9	3.9	18.0	4
.38 Special +P	158 SWC	915	295	280	270	250	215	0.3	1.4	5.6	24.5	4V
.38 Super Auto +P	130 FMC	1245	450	400	365	310	245	0.2	0.8	3.4	15.5	5

.41 Rem. Mag.	210 JSP	1300	790	700	630	525	410	0.2	0.7	3.2	15.0	4V
.44 Rem. Mag.	240 HSP	1350	970	860	765	630	480	0.2	0.7	3.0	14.0	4V
.45 Colt	255 Lead	860	420	400	380	345	285	0.4	1.5	6.5	28.5	5½
.45 Auto	230 FMC	810	335	320	310	285	240	0.4	1.7	7.2	31.0	5

*"V" indicates vented barrel (simulates revolver system with cylinder-to-barrel gap).
NOTE: All of the above data are based on the G_1 Siacci Deceleration Table.

CHAPTER EIGHT

Choosing the Correct Gauge and Load

The selection of the proper gauge and load for the shotshell shooter has become a confusing task. Publication after publication feature articles by "experts" who tout the great ballistic powers of the 20-gauge, 3-inch magnum. Most of these writers hasten to point out the fact that the 20-gauge, 3-inch shell is available with a full 1¼-ounce load and hence, it equals the 12-gauge.

Others feel that the only waterfowl load worth a hoot is the 10-gauge, 3½-inch magnum loaded with BB size shot. And still another will make great claims for the tiny 28-gauge as an upland gun.

The reasons for such claims vary from the simple need for something to write about to the author's own prejudice based on feelings rather than facts.

However, the selection of gauges and loads is not difficult once one becomes familiar with some of the basics.

Gauge selection is very quickly covered. Load selection, however, is another matter. First, let us cover the various gauges and shell lengths.

10-GAUGE 3½" MAGNUM

This shell has only one legitimate application—long-range waterfowl shooting. The monstrous recoil from guns using 10-gauge, 3½-inch shells is seldom mastered. For those few who have extensive shotgunning expertise and who can handle large helpings of abusive recoil, the 10-gauge may have its place. However, this only applies if shots past 50 yards are required in order to take game. Most shooters would be better off to pass the few long-range shots up and use a gun that would enable them to master the shotgunning art.

10-GAUGE 2⅞ "

This length 10-gauge shell offers nothing to the shooter that cannot be accomplished with a 12-gauge gun. Its only real usefulness is that it allows a shooter who realizes that he cannot handle his 3½-inch magnum to shoot a shell that is far less punishing. However, due to lack of popularity, this shell is dying fast.

12-GAUGE 3" MAGNUM

The 12-gauge, 3-inch magnum shotgun is a very versatile gun indeed. Almost all 12-gauge, 3-inch guns will accept the use of standard 12-gauge, 2¾-inch ammunition. The exceptions include some of the automatics that will accept only the high-velocity, 12-gauge, 2¾-inch loads.

Of course, the 3-inch shell offers no advantage for any shotshell shooting other than long-range waterfowl shooting. At that, the heaviest loadings of 1⅞ ounces are loaded only to the modest level of 1210 fps. Most experienced waterfowlers have found that when the temperature drops down around 0° to 10° F. shotshell velocities begin to fall off noticeably. The 1⅞-ounce load then begins to fail to do the job it was designed to do—kill waterfowl cleanly at long ranges. Under these conditions, the lighter 1⅝-ounce load at a muzzle velocity of 1280 fps seems to do the better job.

When one reflects that a standard 2¾-inch, 12-gauge shell will push a 1½-ounce charge at 1260 fps, one is forced to contemplate whether ⅛ ounce of shot and 20 fps can really make it worthwhile to consider the 3-inch shell. The practical answer is, of course, no. It is simply not worth the extra ammunition cost for such a small gain, unless, of course, you are a purist waterfowler. A purist is one who seldom leaves the shoreline for the upland cover, one whose waterfowling is so important that he or she will pay any premium for those extra few yards of range.

The author was such a hunter, and for years a 3-inch gun was the only gun for him when waterfowling. But the middle forties have been somewhat less than kind, and he now does not care for the extra recoil or gun weight. In fact, he really can't stand to shoot more than one box of 3-inch shells without suffering from a mild case of whiplash. He is, however, starting to wonder why he ever went the 3-inch load in the first place. The ducks are still taken just as regularly with 2¾-inch magnum shells as they were with the 3-inch shells. If his effective range has been shortened, he cannot see the difference.

Cutaway of a typical Federal paper Champion 12-gauge target load with a 12C1 wad.

One valid application for the 3-inch shell will always remain—the 1⅞-ounce load of BB size shot for Canada geese. No other load will do as well, nor even come close to doing as well, on Canadas.

12-GAUGE 2¾ "

If there was ever one shotgun shell to do everything, surely the 2¾-inch, 12-gauge is that shotgun shell. The range of loads available can be used successfully for game from the tiny sora rail to the majestic Canada goose or the largest turkey. It can, in fact, be used successfully on deer and black bear. Yes, the 12-gauge has a noticeable amount of recoil with the heaviest loads, but with the 2¾-dram equivalent factory 1⅛-ounce load, recoil is about on par with high-velocity 20-gauge loads. And where further reduction of recoil is required, one can handload 1-ounce loads to a recoil level equal to the lightest 20-gauge loads.

The factory 1-ounce load in 12-gauge actually produces more recoil than the light 1⅛-ounce loads. The 1-ounce load at 1290 fps simply generates more recoil than 1⅛ ounces of shot at 1145 fps. Hence, the 1-ounce factory load seems (due to velocity) to have little value. However, where extremely short-range shots are usual, meat destruction due to the number of pellet hits can be held down with this 1-ounce load.

One of the only reasons for selecting a hunting shotgun of a gauge smaller than 12 should be the weight of the gun. Where a gun is going to be carried all day for upland hunting one can justify the use of a light 20-gauge gun. Of course, if hunting is limited solely to upland birds, squirrels, and rabbits, a gauge smaller than 12 can be used quite successfully.

16-GAUGE 2¾ "

This is, for all practical purposes, a dead gauge. It is not up to 12-gauge performance and offers little or nothing over the 20-gauge. The velocity levels of the 1-ounce load equal the velocity levels of the light 1-ounce, 20-gauge load. There is a 1-ounce, 20-gauge load quite a bit faster (see Chapter Nine).

The 16-gauge, 1⅛-ounce load in standard velocity is only 10 fps faster than the 20-gauge, 1⅛-ounce (2¾-inch magnum). The 1¼-ounce load does have a good edge on the 20-gauge 3-inch 1¼-ounce load, but it is a far cry from the velocities of the 12-gauge.

The choice of a 16-gauge gun today is not a wise one. Gun for gun, there is seldom a quarter of a pound difference in a 16-gauge and a 12-gauge gun. With no real weight advantage over the 12-gauge and a serious ballistic handicap, the 16-gauge simply doesn't make good sense. Very few 16-gauge guns are being produced today and for good reason.

One final comment: It has been the author's experience that 16-gauge ammunition seldom is as fast as the advertised velocities.

20-GAUGE 3"

In a lightweight gun, the 20-gauge, 3-inch shell offers a very versatile upland-game gun that can, in a pinch, double as a duck gun for decoying birds.

The 1¼-ounce, 20-gauge load is a ballistic midget compared to the 1¼-ounce, 12-gauge load (1185 fps vs. 1330 fps), but it will take ducks to 35 or 40 yards under good conditions. It is not, and never will be, an ideal or anywhere near ideal waterfowl gun. It simply doesn't have the velocity to get the job done well. Under most of today's waterfowling conditions a 1¼-ounce load is not really heavy enough for the ranges normally afforded the average gunner. Twelve-gauge loads of 1⅜ ounces or 1½ ounces are the ideal way to go for waterfowl.

The 3-inch, 20-gauge gun, however, does a pretty good job on long-range hunting for pheasants, and it is ideal for most other upland situations.

20-GAUGE 2¾"

A perennial favorite, the 20-gauge, 2¾-inch is frequently abused. In its maximum loading it moves a 1⅛-ounce charge at 1175 fps. While this is the 2¾-inch magnum load for the 20, the same shot charge and velocities are target loads and light upland loads in the 12-gauge. Obviously, the smaller gauge offers no magical powers, and the 2¾-inch, 20-gauge is, at best, a good upland bird, squirrel, or rabbit gauge. It does not even have enough to consistently take pheasants at ranges which exceed 40 yards. Nor would a good sportsman think of it as a deer or bear gauge.

However, the delight of a 20-gauge in woodcock or other upland cover cannot be matched. In this habitat, the 20-gauge will get the job done when the right loads are used. And it will get the job done with a lightweight gun that proves a pleasure to carry and shoot.

The standard 20-gauge is also the correct choice for any shooter who is sensitive to recoil (a reloader, as noted earlier, can tame the 12-gauge to this level).

The heaviest 2¾-inch loads prove to be poor choices for water-fowl hunting except at extremely short ranges and under ideal conditions.

28-GAUGE 2¾"

At very short ranges the 28-gauge is capable of taking *small* upland birds, squirrels, and rabbits. However, if the range runs over 25 yards, a great number of cripples are common. It is at its best on a skeet field rather than in the game field. Yes, there are a number of hunters who succeed in taking fair shares of quail, rabbits, woodcock, and other small upland game. However, the number of clean instant kills they acomplish are limited. Even some of the great shots I have hunted with, more often than not, have had to apply the final blow themselves, or have a good retriever do it.

.410 BORE 3"

The .410 is a stunt gun. It is sufficient for rabbits, squirrels, and small birds up to a maximum of 20 yards. Even at these ranges the average shooter will have a great amount of trouble

For all skeet shooting, No. 9 shot is universally used.

making clean kills. The author used a 3-inch .410 for years on rabbits when he was a teenager and young man. He never killed a rabbit past 20 yards except when two shots were taken. The first shot would invariably roll the rabbit, and it was quickly learned that a second shot, while the rabbit was rolling, was required to accomplish a clean job.

.410 BORE 2½ "

The shell is useless except for making the game of skeet very

Woodcock cover calls for No. 9 shot.

difficult. No conscientious sportsman would ever go afield with this shell.

SPECIFIC LOAD SELECTION

The selection of specific loads within a given gauge and shell length can be difficult unless one approaches the selection with a full knowledge of the three parts of selection required. These parts are: first—shot size; second—shot charge weight; and third—velocity.

SHOT SIZE

The correct shot size can be learned through trial and error, if one has the opportunity to do a great amount of hunting. Otherwise, the choice of shot size can be regulated to simple reference tables. Over the years much to-do has been made on the use of small size shot by a number of gun writers. If one were to believe all he reads, you would decide perhaps that the only shot size needed for all upland shooting and even ducks over decoys was 7½. Actually, the use of small size shot causes more lost cripples than one could imagine.

At one time I raised 500 to 1000 pheasants each year, and was present when most of these birds were shot. Almost everyone,

No. 4 shot, at a minimum of 1¼ ounces, is an ideal waterfowl load, although serious duck hunters seem to prefer 1⅜ ounces.

Long-range waterfowling requires a minimum of 1⅜ ounces of No. 4 shot in a 12-gauge gun.

including myself, used 7½'s. But with almost daily shooting, I soon learned to notice a leg that would drop an inch or so and then quickly be pulled back up. I learned how to spot a bird that was dusted and a great number of other hit indications that go unnoticed by the average gunner. I quickly discovered that a great many "misses" were actually hits where the bird simply gave little or no sign of being shot. Also, when gunning the same areas day after day, the number of dead birds found in the cover was startling.

I remember one particular cockbird that was shot at three times by a companion. He thought he had fired all misses. It appeared that the bird's tail had tucked under ever so slightly, but when he flew a good 400 yards before setting his wings to glide into a wooded area, I was convinced he hadn't been hit. The next morning when working that same wooded area we found the big cockbird—dead. He had flown into the woods with wings set, and he died that way before he ever hit the ground. Over the years we finally settled on No. 4 shot as the ideal pheasant shot size.

Waterfowling is my life with respect to my hunting interest. Some 20 years ago, I decided that any crippled duck on the water was a very difficult bird to kill. I further decided that it took a head shot to get the job done. My base shot size for ducks then was a 1¼-ounce load of 5's or 6's. Through the years it became obvious that No. 4's were the smallest size shot to be used on ducks. Size 2 proved to be sufficient on the late-season big blacks, but it was not needed for any other purpose. But I persisted in using 7½'s and head shots for the cripples when they occurred. My success on finishing cripples was not noteworthy.

About four years ago my oldest son, who was then 14, was doing a high school paper on ducks, their food, habitat, etc. He literally autopsied over 200 birds in one season. At the same time his interest in taxidermy had taken hold. He quickly discovered that 7½'s with a muzzle velocity of 1330 fps would not penetrate the skull of a duck more than 20 or so yards away. That was the end of 7½'s for cripples. Now we use our regular duck loads of 1⅜ ounces or 1½ ounces of No. 4's on the cripples and are enjoying a very high rate of success on finishing cripples, when they occur, with only one or two shots.

The following table for shot selection does narrow down the "paint brush stroke" suggestions made by the ammunition man-

The upland pheasant is best hunted with No. 4 shot, but many hunters are successful with shot sizes as small as 7½.

ufacturers. It is, however, based on 26 years of experience and literally over 6000 birds personally shot, in addition to the first-hand experience of others who shot perhaps an equal number of birds and upland game in my presence.

VARIETY SPEED IN FEET PER SECOND

Variety	50	55	60	65	70	75	80	85	90	95	100
CANVAS-BACK									TO		
GREEN-WING TEAL							TO				
BLUE-WING TEAL							TO				
REDHEAD						TO					
BRANT						TO					
CANADA GOOSE						TO					
GADWALL						TO					
WIDGEON						TO					
PINTAIL				TO							
SPOONBILL		TO									
BLACK DUCK		TO									
MALLARD		TO									

Shot Size	Diameter	Application
BB	.18	Geese, turkey, fox, and similar size game.
2	.15	Geese at short ranges, turkey, and fox. Also large ducks such as blacks, mallards, and white-winged scoters.
4	.13	All duck shooting, small geese, pheasants, turkeys, squirrels, and rabbits.
5	.12	Duck shooting at 30 yards or less for all species. Any practical range for the smaller ducks such as old squaw, bufflehead, and

Shot Size	Diameter	Application
		teal. Also can be used on pheasants, rabbits, and large grouse.
6	.11	Pheasants at modest ranges, grouse, doves, pigeons, partridge, rabbits, squirrels, and crows.
7½	.095	Crows, woodcock, snipe, large rails (clapper), rabbits, doves, pigeons, grouse, quail, partridge, and trap shooting.
8	.09	Quail, small grouse, partridge, woodcock, snipe, large rails (clapper), and trap shooting.
9	.08	Quail, woodcock, snipe, all rails, and skeet shooting.
11	.06	Small rails (sora).

SHOT CHARGE WEIGHTS

The selection of the proper size shot does not unto itself guarantee success. The proper weight of the shot charge is vital to ensure patterns dense enough to obtain clean kills at the ranges being hunted. The following should enable you to combine the *minimum*-required shot weight with pellet sizes and application.

Shot Size	Minimum Shot Weight	Applications
BB	1½ ounces	For all applications suggested for this size shot.
2	1⅜ ounces	For all applications suggested for this size shot.
4	1¼ ounces	For all applications of this size shot except rabbits. For rabbits, shot charges with 1⅛ ounces of No. 4's are sufficient.
5	1¼ ounces	For all applications of this size shot except rabbits. For rabbits, shot charges with 1⅛ ounces of No. 5's are sufficient.
6	1⅛ ounces	For all applications of this size shot.
7½	1 ounce	For all applications of this size shot except crows where 1⅛ ounces should be the minimum used.
8	⅞ ounce	For all applications of this size shot except trap shooting. For trap shooting, a minimum of 1 ounce is suggested (1⅛ ounces are standard).
9	¾ ounce	For all applications of this size shot.
11	¾ ounce	For all applications of this size shot.

VELOCITY

The final step in the selection of a shotshell, after determining the shot size and charge weight, is to select the proper velocity level. To match the factory system of listing a dram equivalent on the shell box to a velocity level, refer to Chapter Nine and the various dram-equivalent listings.

Shot Size	Muzzle Minimum Velocity	Application
BB	1210 fps	For all applications of this size shot.
2	1210 fps	For all applications of this size shot.
4	1260 fps	For all applications of this size shot except rabbits. For rabbits, a minimum muzzle velocity of 1135 fps will do.
5	1185 fps	For all applications of this size shot except rabbits. For rabbits, a minimum muzzle velocity of 1135 fps will do.
6	1165 fps	For all applications of this size shot except rabbits. For rabbits, a minimum muzzle velocity of 1135 fps will do.
7½	1165 fps	For all applications of this size shot except rabbits and trap shooting. For these two purposes, a muzzle velocity minimum would be 1135 fps.
8,9 & 11	1135 fps	For all applications of this size shot.

For further clarification and selection of shotshell loads refer to Chapter Eleven.

STEEL SHOT

The Federal Government now requires the use of steel shot in certain areas for waterfowl hunting. The use of steel shot, its effectiveness on waterfowl, and its potential to damage firearms has been, and will continue to be, a very controversial subject. However, discussions of the general merits, or lack thereof, of steel shot are academic. It is now law that steel shot is required in specific areas. As long as the law stands, we can do nothing but gain some knowledge of the characteristics of steel shot and then select our steel loads wisely.

Perhaps one of the best overviews of steel shot I have seen was in a Federal Cartridge Corporation's news release entitled, "Steel Shot—An Overview." Because of the excellent steel-shot coverage given in this news release, I will quote it directly.

"At the present time, steel shot shells are available only in 12

America's greatest game bird can be successfully hunted with No. 2 on BB size shot.

gauge. In some areas designated for steel, hunters using 10-, 16-, or 20-gauge shotguns may continue to use lead shot loads, as steel shot loads are not currently available in these gauges. Local regulations should be checked in this regard prior to hunting. Steel shot shells in other gauges will very likely become available in the future.

"There has been considerable misunderstanding surrounding the capabilities and limitations of steel shot ammunition. Although similar in many ways to shotshells loaded with lead shot, steel shot loads require specially produced shot, special wads, and special powders.

SHOT: Steel shot is made by cutting low carbon, soft steel wire into short lengths which are formed and ground to the proper

Federal 12-gauge steel shot wad with shot charge.

size. The pellets are then annealed and coated with a rust inhibitor. This process, similar to that used in the manufacture of ball bearings, is much more expensive than the tower drop method used for most lead shot.

"The steel pellets are both hard and light. For example, the hardness of lead shot measured by the Diamond Pyramid Hardness (DPH) test seldom exceeds 30 DPH, while that of steel shot is about 90 DPH. By comparison, air rifle shot (BB's) are about 150 DPH and ball bearings about 270 DPH.

"A steel pellet weighs about 30% less than a lead pellet of similar size. By virtue of their lighter weight, more steel pellets are required per unit of weight than lead pellets. For example, about 134 lead No. 4 pellets weigh one ounce; the same weight of steel shot contains approximately 192 pellets. Perhaps most important, the more numerous steel pellets also occupy a greater volume per unit of weight.

"At present, steel shot is being made in standard sizes. The most common sizes of steel shot are No. 1, 2, and 4, although larger sizes have been loaded experimentally. Smaller sizes of steel shot are not presently being considered.

WADS: Lead shot wads are molded of flexible, low density polyethylene plastic. They usually incorporate a cushion section in the bottom which collapses on firing to reduce the number of deformed pellets in the shot charge. In addition, such wads generally feature a shot pouch which prevents the pellets from scrubbing on the bore surface. Steel shot wads are molded of stiff high density polyethylene plastic and do not incorporate a cushion section on the bottom. Although such a cushion section is ballistically desirable, the space is badly needed because of the greater volume occupied by the steel pellets. The hardness of the steel pellets helps prevent undesirable pellet deformation on firing. Steel shot wads must be much thicker than lead shot wads to prevent the pellets from contacting the bore surface. This is absolutely necessary to prevent barrel scratching and to eliminate or minimize choke expansion previously associated with steel-shot loads. Such wads are more expensive to manufacture than plastic lead-shot wads.

POWDER: Large charges of special smokeless powders must be used to provide the required performance for steel shot ammunition. This raises the cost of such loads and creates additional ballistic problems. For example, such heavy charges of powder must compete for space in an already crowded load. As a result, a trade-off must be made between shot charge volume, and powder charge volume to achieve a balanced load.

SELECTION AND PERFORMANCE

"Steel shot shells should not be judged by lead shot criteria. A 12-gauge, 2¾-inch shell containing a 3¾ dram-equivalent (D.E.) loading with 1¼ ounces of lead shot is a popular waterfowl loading; an equivalent steel-shot load is a 3¾ dram-equivalent with a 1⅛-ounce shot charge. Although the steel load has a lighter shot charge, the number of pellets per shell is actually higher:

Shot Size	Number of Lead Pellets in 1¼ Oz.	Number of Steel Pellets in 1⅛ Oz.
6	282	—
4	167	216
2	109	140
1	—	116

"The 3¾ D.E. 1¼-ounce lead load has a muzzle velocity of about 1330 fps. By contrast, the 3¾ D.E. 1⅛-ounce steel load has a higher muzzle velocity of about 1365 fps. Federal Cartridge Corporation believes such a load to be a good balance for maximum effectiveness. During the previous waterfowl seasons, many users agreed.

"Currently, 12-gauge, 2¾-inch steel loads with 1¼ ounces of shot are being introduced. Because the shot charge is heavier and occupies additional volume, the muzzle velocity must be reduced. These loads are the maximum amount of steel shot that can be put into a 12-gauge, 2¾-inch shell.

COMMENTS

"Steel shot has been criticized as being the potential cause of more crippled birds than will be saved from lead poisoning by the switch to nontoxic shot. This contention remains unproven. However, it will be necessary for hunters to revise their shot size selection and shooting techniques to obtain top performance with steel shot. This will require several years. At that time a better evaluation of the crippling effects of steel shot can be made.

"Hunters using steel shot for the first time should keep these considerations in mind.

1. Be certain to select the proper shot size for the type of hunting anticipated. Remember to use larger sizes of steel shot for equivalent downrange performance.
2. Because steel-shot patterns are denser than lead-shot patterns, hunters should consider the potential benefits of selecting a more open choke when switching to steel.
3. *Steel shot is not as effective as lead shot at extreme ranges. Careful selection of shooting conditions will become more important when using steel shot.*

CONCLUSION

"Steel shot ammunition costs more than lead shot ammunition because of the increased cost of the special shot, wads, and powders necessary. At the present time, steel-shot ammunition costs about the same as 12-gauge, 3-inch magnum 1⅞-ounce shells. In the broad view, however, the cost of ammunition is still a very small percentage of the total cost of a hunt.

"What of the future? Steel shot will be an increasing factor in waterfowl hunting as its required use spreads. Compared to the many years of development behind lead shot shells, the development of non-toxic shot shells has barely begun. In the future, such loads will be more effective and less likely to harm gun barrels as the body of knowledge about such loads increases and improvements are made. This is a small price to pay for protecting our waterfowl resources."

The author expresses his thanks to Federal Cartridge Corporation and Mr. Mike Bussard for supplying the foregoing for inclusion in this book.

AMERICAN STANDARD SIZES

SHOT NAME OR NUMBER	NOMINAL DIAMETER IN INCHES		NOMINAL NUMBER OF PELLETS TO THE OUNCE	
			HARD	SOFT
DUST	.04	.	4665	4565
12	.05	.	2385	2335
11	.06	•	1380	1350
10	.07	•	870	850
9	.08	•	585	570
8	.09	•	410	400
7½	.095	•	345	340
7	.10	•	300	290
6	.11	•	225	220
5	.12	●	175	170
4	.13	●	135	130

NOTE: NOMINAL PELLET COUNTS CALCULATED FROM THEORETICAL DIAMETERS USING 4% ANTIMONIAL CONTENT FOR HARD SHOT AND 0.5% FOR SOFT SHOT.

BUCKSHOT

SHOT NAME OR NUMBER	NOMINAL DIAMETER IN INCHES		NOMINAL NUMBER OF PELLETS TO THE POUND	
			HARD	SOFT
NO. 4 BUCK	.24	●		338
NO. 3 BUCK	.25	●		299
NO.2 BUCK	.27	●		238
NO. 1 BUCK	.30	●		173
NO. 0 BUCK	.32	●		143
NO. 00 BUCK	.33	●		130
NO. 000 BUCK	.36	●		100

NOMINAL PELLET COUNTS FOR SIZES 4 AND SMALLER ROUNDED TO THE NEAREST 5 PELLETS.

SHOT NAME OR NUMBER	NOMINAL DIAMETER IN INCHES		NOMINAL NUMBER OF PELLETS TO THE OUNCE	
			HARD	SOFT
3	.14	●	109	106
2	.15	●	88	87
1	.16	●	73	71
B	.17	●		59
AIR RIFLE	.175	●		55
BB	.18	●		50
BBB	.19	●		43
T	.20	●		36
TT	.21	●		31
F	.22	●		27
FF	.23	●		24

NOTE: NOMINAL PELLET COUNTS CALCULATED FROM THEORETICAL DIAMETERS USING 0.5% ANTIMONIAL CONTENT.

CHAPTER NINE

Exterior Ballistics Tables for Shotshells

As with all the previous ballistics tables in the book, the shotshell ballistics tables should not be considered as absolute. Variations in shotguns and ammunition components can create varying ballistics.

The tables on these pages are based on nominal numbers and carried out by computer. They should represent the average results you could expect to obtain in a given string of shots. However, variations of as much as plus or minus 45 fps in muzzle velocity are not uncommon in shotshells. Such changes would, of course, affect all downrange ballistics.

Shotshell ballistics are, of course, affected by the length of the barrel used. The data in the following tables was based on these barrel lengths:

Buckshot and Slugs
12-gauge = 30" barrels
16-gauge = 28" barrels
20-gauge = 26" barrels
28-gauge = 26" barrels
.410 bore = 26" barrels

Normally shotshells are not advertised by a velocity level but rather by a dram-equivalent rating. In order to use the ballistics tables, you will have to convert these dram-equivalent ratings to a velocity. To do so, simply refer to the dram-equivalent rating marked on your box of shells and find that rating under the proper gauge as follows.

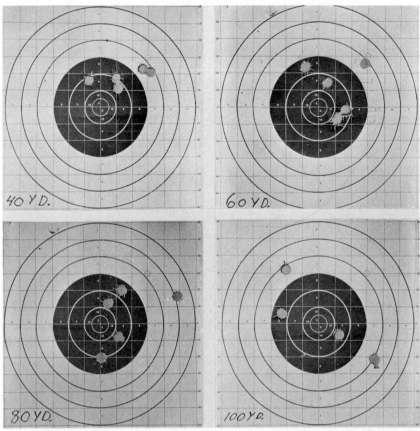

The 12-gauge slug is an excellent deer load to approximately 100 yards when used in a gun properly equipped with good sights. These targets were shot with an Ithaca 37 Deerslayer equipped with a 2½X Lyman scope.

Dram Equivalents
12-gauge (30″ barrel)
2¾″ Shells
3¼ drams equivalent, 1 ounce shot = 1290 fps
2¾ drams equivalent, 1⅛ ounces shot = 1145 fps
3 drams equivalent, 1⅛ ounces shot = 1200 fps
3¼ drams equivalent, 1⅛ ounces shot = 1255 fps
3¼ drams equivalent, 1¼ ounces shot = 1220 fps
3¾ drams equivalent, 1¼ ounces shot = 1330 fps
3¾ drams equivalent, 1½ ounces shot = 1260 fps
3″ Shells
3¾ drams equivalent, 1⅜ ounces shot = 1295 fps
4 drams equivalent, 1⅝ ounces shot = 1280fps
4 drams equivalent, 1⅞ ounces shot = 1210 fps

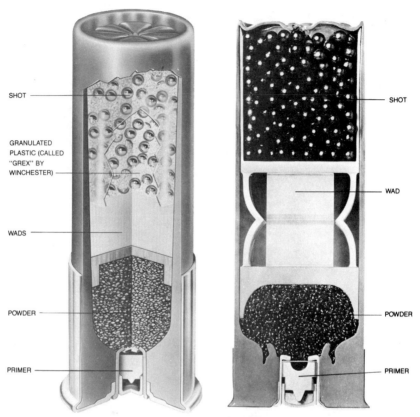

SHOT

GRANULATED PLASTIC (CALLED "GREX" BY WINCHESTER)

WADS

POWDER

PRIMER

SHOT

WAD

POWDER

PRIMER

Left: *Cutaway view of a Winchester-Western Super X Double X shotshell.* Right: *Cutaway view of a Winchester-Western 12-gauge Double A target shell.*

16-gauge (28″ barrel)
2¾″ Shells
2½ drams equivalent, 1 ounce shot = 1165 fps
2¾ drams equivalent, 1⅛ ounces shot = 1185 fps
3¼ drams equivalent, 1⅛ ounces shot = 1295 fps
3¼ drams equivalent, 1¼ ounces shot = 1260 fps
20-gauge (26″ barrel)
2¾″ Shells
2½ drams equivalent, ⅞ ounce shot = 1210 fps
2½ drams equivalent, 1 ounce shot = 1165 fps
2¾ drams equivalent, 1 ounce shot = 1220 fps
2¾ drams equivalent, 1⅛ ounces shot = 1175 fps
3″ Shells
3½ drams equivalent, 1³⁄₁₆ ounces shot = 1295 fps
3 drams equivalent, 1¼ ounces shot = 1185 fps

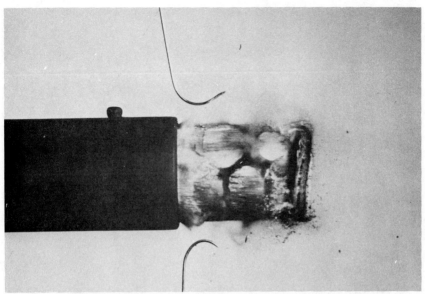

Standard 12-gauge .00 buckshot load—12 pellets at the muzzle.

Rifle slug leaving the muzzle of a 12-gauge shotgun.

28-gauge (26″ barrel)
2¾″ Shells
2¼ drams equivalent, ¾ ounce shot = 1295 fps
.410 Bore (26″ barrel)
2½″ Shells
Max., ½ ounce shot = 1135 fps
3″ Shells
Max., $^{11}\!/_{16}$ ounce shot = 1135 fps

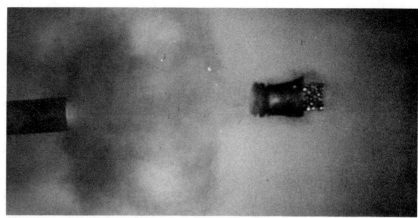

Shot charge leaving the muzzle of a shotgun.

SHOTSHELL VELOCITY vs. BARREL LENGTH

Buckshot & Rifled Slug Loads

Gauge	Nominal Barrel Length	Range of Barrel Lengths to Which Data Applies (inches)	Loss or Increase in Velocity for each 2" Shorter or Longer than Nominal Barrel (fps)
10	30"	32-22	15
12	30"	32-22	15
16	28"	30-20	15
20	26"	30-20	15
28	26"	28-20	15
.410 (bore)	26"	28-20	25

Bird-Shot Loads

Gauge	Nominal Barrel Length	Range of Barrel Lengths to Which Data Applies (inches)	Loss or Increase in Velocity for each 2" Shorter or Longer than Nominal Barrel (fps)
10	30"	32-22	10
12	30"	32-22	10
16	28"	30-20	15
20	26"	30-20	15
28	26"	28-20	25
.410 (bore)	26"	28-20	25

EXTERIOR BALLISTICS TABLES—SHOTGUN

Rifled Slugs

Gauge	Wgt. Oz.	Velocity in fps at					Energy in ft.-lbs. at				
		Muzzle	25 yds.	50 yds.	75 yds.	100 yds.	Muzzle	25 yds.	50 yds.	75 yds.	100 yds.
12	7/8	1600	1365	1175	1040	950	2177	1584	1174	920	767
16	4/5	1600	1365	1175	1040	950	1990	1448	1073	840	701
20	5/8	1600	1365	1175	1040	950	1555	1130	840	655	550
.410	1/5	1830	1560	1335	1150	1025	650	475	345	255	205

EXTERIOR BALLISTICS TABLES—SHOTGUN

Buckshot

Gauge	Buck Size	Velocity in fps at							Energy in ft.-lbs. per Pellet at						
		Muzzle	10 yds.	20 yds.	30 yds.	40 yds.	50 yds.	60 yds.	Muzzle	10 yds.	20 yds.	30 yds.	40 yds.	50 yds.	60 yds.
12	00	1325	1220	1135	1070	1015	970	930	210	180	155	135	125	110	105
12	00	1250	1160	1085	1030	985	940	900	185	160	140	125	115	105	95
12	0	1300	1200	1120	1055	1005	960	915	185	160	135	120	110	100	90
12	1	1250	1135	1050	990	935	885	835	140	115	100	90	80	70	65
12	4	1325	1195	1095	1020	960	905	860	80	65	55	50	45	40	35
16	1	1225	1115	1040	975	925	875	830	135	110	95	85	75	70	60
20	3	1200	1100	1025	970	915	870	825	75	65	55	50	45	40	35

EXTERIOR BALLISTICS TABLES—SHOTGUN

Rifled Slugs

Gauge	Slug Wgt. Oz.	Vel. fps at Muzzle	Time of Flight in Sec. to				Drop in Inches at				Midrange Trajectory in In. for a Range of			
			25 yds.	50 yds.	75 yds.	100 yds.	25 yds.	50 yds.	75 yds.	100 yds.	25 yds.	50 yds.	75 yds.	100 yds.
12	7/8	1600	.051	.110	.178	.254	.5	2.1	5.3	10.4	.1	.6	1.5	3.1
16	4/5	1600	.051	.110	.178	.254	.5	2.1	5.3	10.4	.1	.6	1.5	3.1
20	5/8	1600	.051	.110	.178	.354	.5	2.1	5.3	10.4	.1	.6	1.5	3.1
410	1/5	1830	.044	.096	.157	.226	.4	1.6	4.1	8.2	.1	.4	1.2	2.5

Buckshot

Gauge	Size	Vel. fps at Muzzle	Time of Flight in Sec. to						Drop in Inches at		Midrange Trajectory in In. for a Range of			
			10 yds.	20 yds.	30 yds.	40 yds.	50 yds.	60 yds.	10 yds.	20 yds.	30 yds.	40 yds.	50 yds.	60 yds.
12	00	1325	.024	.049	.076	.105	.135	.167	.1	.4	1.0	2.0	3.2	4.8
12	00	1250	.025	.052	.080	.110	.141	.174	.1	.5	1.2	2.2	3.5	5.2
12	0	1300	.024	.050	.078	.107	.137	.169	.1	.5	1.1	2.0	3.3	4.9
12	1	1250	.030	.063	.099	.136	.176	.217	.2	.7	1.7	3.2	5.2	7.8
12	4	1325	.024	.050	.079	.109	.141	.175	.1	.5	1.1	2.1	3.4	5.1
16	1	1225	.031	.064	.100	.138	.178	.220	.2	.7	1.8	3.3	5.4	8.0
20	3	1200	.026	.054	.085	.116	.150	.186	.1	.5	1.3	2.4	3.9	5.9

EXTERIOR BALLISTICS TABLES—SHOTGUN

Velocity in fps at

Shot Size	Muzzle	10 yds.	20 yds.	30 yds.	40 yds.	50 yds.	60 yds.
4	1360	1170	1030	915	825	755	690
5	1360	1160	1010	890	800	725	665
6	1360	1145	985	865	770	695	635
BB	1330	1195	1085	990	915	850	790
2	1330	1170	1045	945	860	790	730
4	1330	1150	1010	905	815	745	685
5	1330	1135	990	880	790	715	655
6	1330	1120	970	855	765	690	630
7½	1330	1095	930	810	715	640	580
BB	1315	1180	1075	985	905	840	785
2	1315	1160	1035	935	855	785	725
4	1315	1140	1005	895	810	740	680
5	1315	1125	985	875	785	715	655
6	1315	1110	960	850	760	685	625
2	1295	1145	1025	925	845	780	720
4	1295	1125	990	885	800	735	675
5	1295	1110	970	865	780	710	650
6	1295	1095	950	840	750	680	620
7½	1295	1070	910	795	705	630	575
BB	1255	1135	1035	950	880	820	765
2	1255	1110	995	905	830	765	705
4	1255	1090	965	870	785	720	665
5	1255	1080	950	845	765	695	640
6	1255	1070	930	820	740	670	610
8	1255	1035	880	765	675	605	550
2	1240	1100	990	900	820	760	705
4	1240	1080	960	860	780	715	660
5	1240	1070	940	840	760	690	635
6	1240	1055	920	815	730	665	610
7½	1240	1035	885	770	690	620	560
4	1235	1075	955	860	780	715	660
5	1235	1065	940	835	755	690	635
6	1235	1055	920	815	730	665	605
8	1235	1020	870	755	670	600	545

Energy in ft.-lbs. per Pellet at

Shot Size	Muzzle	10 yds.	20 yds.	30 yds.	40 yds.	50 yds.	60 yds.
4	13.28	9.85	7.60	6.03	4.91	4.07	3.43
5	10.54	7.64	5.79	4.54	3.65	3.00	2.51
6	7.95	5.62	4.17	3.22	2.56	2.09	1.74
BB	34.37	27.75	22.87	19.04	16.27	14.04	12.23
2	19.07	14.76	11.77	9.60	7.98	6.74	5.76
4	12.70	9.50	7.34	5.85	4.77	3.97	3.35
5	10.08	7.36	5.60	4.41	3.56	2.93	2.46
6	7.61	5.41	4.04	3.13	2.50	2.04	1.70
7½	4.88	3.30	2.38	1.80	1.41	1.13	0.93
BB	33.60	27.05	22.45	18.85	15.91	13.71	11.97
2	18.64	14.47	11.56	9.45	7.86	6.65	5.69
4	12.42	9.29	7.22	5.77	4.71	3.92	3.32
5	9.86	7.22	5.51	4.34	3.51	2.90	2.43
6	7.44	5.31	3.98	3.09	2.47	2.02	1.68
2	18.08	14.09	11.28	9.24	7.71	6.53	5.60
4	12.04	9.05	7.05	5.65	4.62	3.86	3.26
5	9.56	7.03	5.38	4.25	3.45	2.85	2.40
6	7.17	5.17	3.89	3.03	2.43	1.99	1.66
7½	4.63	3.16	2.30	1.74	1.37	1.10	0.91
BB	30.60	25.03	20.81	17.54	15.05	13.06	11.37
2	16.98	13.25	10.67	8.83	7.43	6.31	5.86
4	11.31	8.57	6.72	5.41	4.45	3.72	3.16
5	8.96	6.66	5.13	4.08	3.32	2.75	2.32
6	6.77	4.91	3.71	2.91	2.34	1.92	1.61
8	3.69	2.50	1.80	1.36	1.07	0.86	0.70
2	16.58	13.05	10.53	8.68	7.28	6.19	5.33
4	11.04	8.39	6.59	5.32	4.38	3.67	3.12
5	8.76	6.52	5.04	4.01	3.27	2.72	2.29
6	6.61	4.80	3.65	2.86	2.30	1.90	1.58
7½	4.24	2.94	2.16	1.64	1.30	1.06	0.87
4	10.95	8.33	6.55	5.29	4.36	3.65	3.10
5	8.59	6.48	5.01	3.99	3.25	2.70	2.28
6	6.56	4.77	3.62	2.85	2.29	1.89	1.58
8	3.57	2.43	1.76	1.33	1.05	0.84	0.69

EXTERIOR BALLISTICS TABLES—SHOTGUN

Velocity in fps at

Shot Size	Muzzle	10 yds.	20 yds.	30 yds.	40 yds.	50 yds.	60 yds.
2	1220	1085	975	885	815	750	695
4	1220	1065	945	850	775	710	655
5	1220	1055	930	830	750	685	630
6	1220	1040	910	805	725	660	605
7½	1220	1020	875	765	680	615	560
8	1220	1010	860	750	665	595	540
4	1200	1050	935	840	765	700	650
5	1200	1040	915	820	740	680	625
6	1200	1030	900	800	720	655	600
7½	1200	1005	865	760	675	610	555
8	1200	995	850	745	660	590	540
9	1200	975	820	710	625	555	505
4	1185	1040	925	835	760	695	645
6	1185	1015	890	790	715	650	595
8	1185	985	845	735	655	590	535
4	1165	1025	915	825	750	690	635
5	1165	1015	985	805	730	665	615
6	1165	1000	880	785	705	640	590
8	1165	970	835	730	650	585	530
9	1165	950	805	695	615	550	495
4	1155	1015	905	820	745	685	635
5	1155	1005	890	800	725	665	615
6	1155	995	875	780	700	640	485
8	1155	965	830	725	645	580	530
9	1155	945	800	695	610	545	495
9	1150	940	800	690	610	545	495
7½	1145	965	835	735	655	595	540
8	1145	960	825	720	640	580	525
4	1135	1000	895	810	740	680	630
5	1135	990	880	790	715	650	605
6	1135	980	860	770	695	635	580
7½	1135	960	830	730	655	590	540

Energy in ft.-lbs. per Pellet at

Shot Size	Muzzle	10 yds.	20 yds.	30 yds.	40 yds.	50 yds.	60 yds.
2	16.04	12.67	10.26	8.48	7.13	6.07	5.23
4	10.69	8.17	6.43	5.19	4.29	3.60	3.06
5	8.48	5.34	4.92	3.93	3.21	2.67	2.25
6	6.40	4.67	3.56	2.80	2.26	1.86	1.59
7½	4.11	2.86	2.11	1.62	1.28	1.04	0.86
8	3.48	2.38	1.73	1.32	1.03	0.83	0.69
4	10.34	7.92	6.27	5.08	4.20	3.53	3.01
5	8.21	6.16	4.80	3.84	3.14	2.62	2.22
6	6.19	4.54	3.47	2.74	2.22	1.83	1.54
7½	3.97	2.79	2.06	1.59	1.26	1.02	0.85
8	3.37	2.32	1.69	1.29	1.02	0.82	0.68
9	2.38	1.57	1.11	0.83	0.64	0.51	0.42
4	10.08	7.75	6.15	4.99	4.13	3.48	2.97
6	6.04	4.44	3.44	2.69	2.18	1.81	1.52
8	3.29	2.27	1.66	1.27	1.00	0.81	0.67
4	9.74	7.52	5.98	4.87	4.04	3.41	2.91
5	7.74	5.85	4.58	3.69	3.03	2.53	2.15
6	5.84	4.32	3.32	2.63	2.14	1.77	1.49
8	3.18	2.21	1.62	1.24	0.98	0.80	0.66
9	2.24	1.49	1.07	0.80	0.62	0.50	0.41
4	9.58	7.41	5.90	4.81	4.00	3.38	2.89
5	7.60	5.77	4.52	3.64	3.00	2.51	2.17
6	5.74	4.25	3.28	2.60	2.12	1.76	1.48
8	3.18	2.18	1.60	1.23	0.97	0.79	0.65
9	2.20	1.47	1.06	0.79	0.62	0.49	0.40
9	2.18	1.46	1.05	0.79	0.61	0.49	0.40
7½	3.62	2.57	1.93	1.49	1.19	0.97	0.81
8	3.07	2.15	1.58	1.22	0.96	0.78	0.65
4	9.25	7.19	6.74	4.70	3.91	3.31	2.83
5	7.34	5.59	4.40	3.55	2.93	2.46	2.09
6	5.54	4.13	3.19	2.54	2.07	1.72	1.45
7½	3.56	2.54	1.90	1.48	1.18	0.97	0.80

EXTERIOR BALLISTICS TABLES—SHOTGUN

Shot Size	fps at Muzzle	Time of Flight in Sec. to						Drop in Inches at					
		10 yds.	20 yds.	30 yds.	40 yds.	50 yds.	60 yds.	10 yds.	20 yds.	30 yds.	40 yds.	50 yds.	60 yds.
4	1360	.024	.051	.082	.117	.155	.196	0.1	0.5	1.3	2.6	4.6	7.4
5	1360	.024	.052	.084	.119	.158	.202	0.1	0.5	1.4	2.7	4.8	7.8
6	1360	.024	.053	.085	.122	.163	.208	0.1	0.5	1.4	2.9	5.1	8.3
BB	1330	.024	.050	.079	.111	.145	.182	0.1	0.5	1.2	2.4	4.0	6.4
2	1330	.024	.051	.082	.115	.151	.191	0.1	0.5	1.3	2.6	4.4	7.0
4	1330	.024	.052	.084	.119	.157	.199	0.1	0.5	1.4	2.7	4.8	7.7
5	1330	.025	.053	.085	.121	.161	.205	0.1	0.5	1.4	2.8	5.0	8.1
6	1330	.025	.054	.087	.124	.165	.211	0.1	0.6	1.4	3.0	5.3	8.6
7½	1330	.025	.055	.090	.129	.174	.223	0.1	0.6	1.5	3.2	5.8	9.6
BB	1315	.024	.051	.080	.112	.146	.183	0.1	0.5	1.2	2.4	4.1	6.5
2	1315	.024	.052	.082	.116	.153	.192	0.1	0.5	1.3	2.6	4.5	7.1
4	1315	.025	.053	.084	.120	.159	.201	0.1	0.5	1.4	2.9	4.8	7.8
5	1315	.026	.053	.086	.122	.162	.206	0.1	0.6	1.4	2.9	5.1	8.2
6	1315	.026	.054	.087	.125	.167	.212	0.1	0.6	1.5	3.0	5.4	8.7
2	1295	.025	.053	.083	.117	.154	.194	0.1	0.5	1.3	2.6	4.0	7.8
4	1295	.025	.053	.086	.121	.160	.203	0.1	0.6	1.4	2.8	5.0	8.0
5	1295	.025	.054	.087	.124	.164	.208	0.1	0.6	1.4	2.9	5.2	8.4
6	1295	.026	.055	.088	.126	.168	.215	0.1	0.6	1.5	3.1	5.5	8.9
7½	1295	.026	.056	.091	.132	.177	.227	0.1	0.6	1.6	3.3	6.0	9.0
BB	1255	.025	.053	.083	.116	.151	.189	0.1	0.5	1.3	2.6	4.4	6.9
2	1255	.026	.054	.086	.120	.158	.199	0.1	0.6	1.4	2.8	4.8	7.7
4	1255	.026	.055	.088	.124	.164	.207	0.1	0.6	1.5	3.0	5.2	8.3
5	1255	.026	.056	.089	.126	.168	.213	0.1	0.6	1.5	3.1	5.4	8.7
6	1255	.026	.056	.091	.129	.172	.219	0.1	0.6	1.6	3.2	5.7	9.2
8	1255	.027	.058	.095	.137	.184	.236	0.1	0.6	1.7	3.6	6.5	10.7
2	1240	.026	.055	.086	.121	.159	.201	0.1	0.6	1.4	2.8	4.9	7.8
4	1240	.026	.056	.089	.125	.165	.209	0.1	0.6	1.5	3.0	5.3	8.4
5	1240	.026	.056	.090	.128	.169	.215	0.1	0.6	1.6	3.1	5.5	8.9
6	1240	.026	.057	.091	.130	.173	.221	0.1	0.6	1.6	3.3	5.8	9.4
7½	1240	.027	.058	.094	.136	.184	.233	0.1	0.6	1.7	3.6	6.5	10.5
4	1235	.026	.056	.089	.127	.166	.210	0.1	0.6	1.7	3.0	5.3	8.5
5	1235	.026	.056	.090	.128	.170	.215	0.1	0.6	1.8	3.2	5.6	8.9
6	1235	.026	.057	.092	.131	.174	.221	0.1	0.6	1.8	3.3	5.8	9.4
8	1235	.027	.059	.099	.138	.186	.238	0.1	0.7	2.1	3.7	6.6	11.0

EXTERIOR BALLISTICS TABLES—SHOTGUN

Size	Muzzle	Time of Flight in Seconds to						Drop in Inches at					
		10 yds.	20 yds.	30 yds.	40 yds.	50 yds.	60 yds.	10 yds.	20 yds.	30 yds.	40 yds.	50 yds.	60 yds.
2	1220	.026	.055	.088	.123	.161	.203	0.1	0.6	1.5	2.9	5.0	8.0
4	1220	.026	.056	.090	.127	.167	.212	0.1	0.6	1.6	3.1	5.4	8.6
5	1220	.027	.057	.091	.129	.171	.217	0.1	0.6	1.6	3.2	5.6	9.1
6	1220	.027	.058	.093	.132	.175	.223	0.1	0.6	1.7	3.6	5.9	9.6
7½	1220	.027	.059	.096	.137	.184	.235	0.1	0.7	1.8	3.6	5.6	10.2
8	1220	.027	.059	.097	.139	.187	.240	0.1	0.7	1.8	3.8	6.8	11.1
4	1200	.027	.057	.091	.128	.168	.214	0.1	0.6	1.6	3.2	5.4	8.8
5	1200	.027	.058	.092	.131	.173	.219	0.1	0.6	1.6	3.3	5.8	9.3
6	1200	.027	.058	.094	.137	.177	.226	0.1	0.7	1.7	3.4	6.1	9.8
7½	1200	.027	.060	.097	.139	.186	.238	0.1	0.7	1.8	3.7	6.7	10.9
8	1200	.027	.060	.098	.141	.189	.242	0.1	0.7	1.9	3.8	6.9	11.3
9	1200	.028	.062	.101	.146	.197	.254	0.1	0.7	2.0	4.1	7.5	12.4
4	1185	.027	.058	.092	.130	.171	.216	0.1	0.6	1.6	3.2	5.6	9.0
6	1185	.027	.059	.095	.135	.179	.227	0.1	0.7	1.7	3.5	6.2	10.0
8	1185	.028	.061	.099	.142	.191	.244	0.2	0.7	1.9	3.9	7.0	11.5
4	1165	.028	.059	.093	.131	.173	.219	0.2	0.7	1.7	3.3	5.8	9.2
5	1165	.028	.059	.095	.134	.179	.224	0.2	0.7	1.7	3.5	6.0	9.7
6	1165	.028	.060	.096	.137	.181	.230	0.2	0.7	1.8	3.6	6.3	10.2
8	1165	.028	.062	.100	.144	.193	.247	0.2	0.7	1.9	4.0	7.2	11.8
9	1165	.029	.063	.103	.149	.201	.258	0.2	0.8	2.1	4.3	7.8	12.9
4	1155	.028	.059	.094	.132	.174	.220	0.2	0.7	1.7	3.4	5.9	9.3
5	1155	.028	.060	.095	.135	.178	.225	0.2	0.7	1.8	3.1	6.1	9.8
6	1155	.028	.060	.097	.137	.182	.231	0.2	0.7	1.8	3.6	6.4	10.3
8	1155	.029	.062	.101	.145	.194	.248	0.2	0.8	2.0	4.0	7.3	11.9
9	1155	.029	.064	.104	.150	.202	.260	0.2	0.8	2.1	4.3	7.9	13.0
9	1150	.029	.064	.104	.151	.203	.260	0.2	0.8	2.1	4.4	7.9	13.1
7½	1145	.029	.062	.101	.144	.192	.245	0.2	0.7	2.0	4.0	7.1	11.6
8	1145	.029	.063	.102	.146	.195	.250	0.2	0.8	2.1	4.1	7.4	12.0
4	1135	.028	.060	.095	.134	.177	.223	0.2	0.7	1.8	3.5	6.0	9.6
5	1135	.028	.061	.097	.137	.180	.228	0.2	0.7	1.8	3.6	6.3	10.0
6	1135	.029	.061	.098	.139	.185	.234	0.2	0.7	1.9	3.7	6.6	10.6
7½	1135	.029	.063	.101	.145	.193	.246	0.2	0.8	2.0	4.0	7.2	11.7

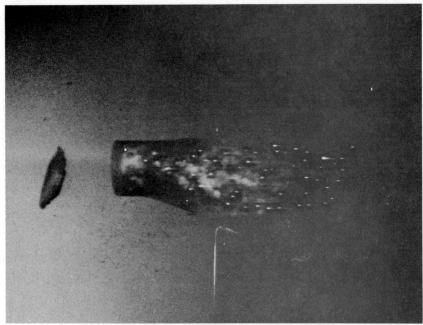

Shot charge with wad starting to fall away.

EXTERIOR BALLISTICS TABLES—SHOTGUN

Approximate Horizontal Distance in Yards
to Point of First Impact for Leading Pellets of the Shot String

Muzzle Velocity*

Shot Size	1350 fps Distance in Yards	1275 fps Distance in Yards	1200 fps Distance in Yards	1125 fps Distance in Yards
00 Bk	610	604	597	587
0 Bk	590	580	574	567
1 Bk	567	557	550	544
2 Bk	527	517	510	504
3 Bk	497	490	484	477
4 Bk	480	474	467	460

Approximate Maximum Horizontal Distance in Yards
to Point of First Impact for Rifled Slugs

Gauge	Slug Wgt.	Max. Range Yards	Muzzle Vel. fps
12	7/8 oz.	817	1600
16	4/5 oz.	817	1600
20	5/8 oz.	817	1600
410	1/5 oz.	844	1830

*Values for intermediate velocities to be obtained by interpolation.

EXTERIOR BALLISTICS TABLES—SHOTGUN

Approximate Maximum Horizontal Distance in Yards
to Point of First Impact
for Leading Pellets**
of the Shot String

Muzzle Velocity*

Shot Size	1350 fps Distance in Yards	1275 fps Distance in Yards	1200 fps Distance in Yards	1125 fps Distance in Yards
FF	467	460	454	447
F	450	444	437	430
TT	434	427	420	414
T	420	414	407	400
BBB	404	397	390	384
BB	387	380	374	367
B	370	364	357	354
1	354	347	344	337
2	337	330	327	320
3	320	317	310	304
4	304	300	294	287
5	290	284	277	274
6	274	267	260	257
7	257	250	247	240
7½	247	244	237	234
8	240	234	230	224
9	224	217	214	207

*Values for intermediate velocities to be obtained by interpolation.

**No allowance for balled shot.

CHAPTER TEN

The Do's and Don'ts of Ammunition Interchangeability

Most of us, probably since the time we were kids, have been aware of cartridge interchangeability within a given firearm. After all, didn't we load up our .22 Long rifle with Shorts because they were considerably less expensive to burn up when engaged in the serious business of busting up a row of tin cans. And why shouldn't we have? The barrel was clearly labeled "chambered for: .22 Shorts, Longs, and Long Rifles." Yes, there were a few exceptions—when a gun chambered for a .22 Long Rifle would not properly function with the shorter rounds. But these firearms were clearly marked "for .22 Long Rifle cartridges only."

Our early habits tend to stay with us. Shooters are constantly concerned with interchangeability, especially when a savings is possible for practice shooting. Some shooters seek lighter recoiling loads, most noticeably perhaps in the big magnum handguns. This may cause problems. Just because a cartridge fits into a chamber does not ensure that it is interchangeable with the cartridge inscribed on the gun. In fact, more often than not, such "fits" are usually potentially dangerous. Some of these "fits" can cause serious damage to the firearm and, in some cases, injure the shooter severely.

High-velocity cartridges should not be used in firearms manufactured prior to the introduction of such cartridges. When in doubt, consult the manufacturer.

The use of short cartridges in long chambers tends to cause erosion of the chamber. With continued use, such erosion can cause possible difficult extraction and, in extreme cases, ruin the chamber for further use of the longer cartridge.

The following cartridges are those that can be interchanged

safely. No other cartridge switching should ever be attempted. Further, no converse interchangeability should be attempted.

All Rimfire Interchangeability

Firearm marked for:	Can also be used with:
.22 Short	.22 BB Cap, .22 CB Cap, .22 Short Blank, 22 CB Short.
.22 Long	.22 BB Cap, .22 CB Cap, .22 Short Blank, .22 Short, .22 CB Short, .22 CB Long.
.22 Long Rifle	.22 BB Cap, .22 CB Cap, .22 Short Blank, .22 CB Short, .22 CB Long .22 Long, .22 L.R. Shot.
.22 Win. Mag. R.F.	.22 Win. R.F., .22 Rem. Spl.
.22 Rem. Spl.	.22 Win. R.F.
.22 Win. R.F.	.22 Rem. Spl.
.25 Stevens	.25 Stevens Short
.32 Long	.32 Short

The use of short cartridges in guns marked such as "for use with .22 Long Rifle ammunition only" is of course to be avoided. The old but fine Model 63 Winchester Semi-Automatic .22 is an example of such a condition.

Centerfire Handgun Interchangeability

Firearm marked for:	Can also be used with:
.32 Smith & Wesson Long	.32 Smith & Wesson, .32 Smith & Wesson Blank, .32 Colt New Police
.32 Colt New Police	.32 Smith & Wesson, .32 Smith & Wesson Blank, .32 Smith & Wesson Long
.32 Long Colt	.32 Short Colt
.38 Long Colt	.38 Short Colt
.38 Smith & Wesson	.38 Colt New Police, .38 Smith & Wesson Blank
.38 Colt New Police	.38 Smith & Wesson, .38 Smith & Wesson Blank
.38 Special	.38 Short Colt, .38 Long Colt, .38 Special Blank
.357 Magnum	.38 Short Colt, .38 Long Colt, .38 Special Blank, .38 Special, .38 Special +P
.38-40 Winchester	5 in 1 Blank
.38 Super Auto	.38 Auto Colt
.44 S&W Special	.44 S&W Russian
.44 Remington	.44 S&W Special

Magnum
.44-40 Winchester   5 in 1 blank
.45 Colt   5 in 1 blank

The .38 Special +P ammo should only be used in .38-caliber guns that have been designated or recommended by the firearm manufacturer as suitable for such use.

CENTERFIRE RIFLE INTERCHANGEABILITY

The only interchangeability that exists in this area is that guns chambered for the .45-90 Winchester can be used with .45-70 Government. Chamber erosion could occur with this short round and cause difficult extraction with the longer shell. Due to the unavailability of the longer shells, this is probably of no consequence.

Often other centerfire combinations are spoken of (and perhaps correctly so) as being interchangeable. However, upon examination of such combinations, we find that different nomenclature was used for the same cartridge. Such confusion in nomenclature resulted from many reasons, but primarily it was the result of firearms manufacturers, marketing policies. These interchangeable names are as follows:

Centerfire Handguns

Full name	*also called*
.25 Automatic·	.25 Auto, .25 ACP, .25 C.A.P., 6.35mm Auto, 635mm Browning (Auto)
.30 Luger	7.65 Luger, 7.65 Parabellum
.32 Automatic	.32 Auto, .32 ACP, .32 C.A.P., 7.65 Auto, 7.65mm Browning (Auto)
9mm Luger	9mm Parabellum
.380 Automatic	9mm Corto, 9mm Kurtz
.38-40 Winchester	.38-40, .38 W.C.F., .38 Winchester, .38-40 Remington, .38-40 Marlin
.44-40 Winchester	.44-40, .44 W.C.F., .44 Winchester, .44-40 Remington, .44-40 Marlin

CENTERFIRE RIFLES

Full name	*also called*
6mm Remington	(formerly) .244 Remington
.25-20 Winchester	.25-20, .25 W.C.F., .25-20 Marlin
.30-30 Winchester	.30-30, .30 Winchester, .30 Marlin, .30 Savage, .30 W.C.F.

.32-20 Winchester	.32-20, .32 Winchester, .32 Marlin, .32 Remington, .32 W.C.F., .32 Colt L.M.R.
.38-40 Winchester	.38-40, .38 W.C.F., .38 Winchester, .38-40 Remington, .38-40 Marlin.
.44-40 Winchester	.44-40, .44 W.C.F., .44 Winchester, .44-40 Remington, .44-40 Marlin.
.45-70 Government	.45-70, .45-70 Marlin, .45-70-405, .45-70-500

SHOTSHELL INTERCHANGEABILITY

Interchangeability in shotgun shells is limited strictly to the use of short shells in long chambers. This means that 10-gauge, 2⅞-inch shells can be used in 10-gauge, 3½-inch chambers; 12-gauge 2¾-inch shells can be used in 12-gauge, 3-inch chambers; 16-gauge, 2⁹⁄₁₆-inch shells can be used in 16-gauge, 2¾-inch chambers; 20-gauge, 2¾-inch shells can be used in 20-gauge, 3-inch chambers; and finally, .410 bore, 2½-inch shells can be used in .410 bore, 3-inch chambers.

Caution: One of the real problems in interchanging shells occurs when some neophyte or overzealous clerk "discovers" that a cartridge or shell of one type fits a gun of another chamber type. The following list attempts to show some of these so-called "discoveries," all of which create potentially dangerous conditions. The author makes no claim that this list is complete. It merely represents those that we are aware of at this time. It further will serve as WARNING to those shooters who own guns of the calibers involved with potentially dangerous situations.

If you own guns of such calibers it pays to be doubly certain that you have not grabbed the wrong ammo. Be sure to check if you accidentally have left a few rounds of one caliber or gauge ammunition in your pockets before loading up with a new batch of shells for a gun of the other caliber. Be especially careful on outings to the range where dangerous combinations of ammo and firearms may be on hand simultaneously.

Except as earlier noted, a firearm should be used only with the ammunition for which it was originally designed and intended. The combinations in the following listing are typical of those in which a cartridge of one caliber will generally fire in a firearm of another caliber. Such errors can result in split or ruptured cartridge cases. Or they could lead to possible serious injury to either the shooter or bystanders. The possibility of leaving a bullet stuck in the bore, thus forming an obstruction, also exists.

DANGEROUS ARMS AND AMMUNITION COMBINATIONS

Centerfire Handguns

Chambered for:	*Dangerous ammo, applications*
.32 Smith & Wesson	.32 Automatic
	.32 Short Colt
	.32 Long Colt
.38 Automatic	.38 Super Automatic
	.38 Super Automatic +P
.32-20 Winchester	.32-20 High Velocity
.38 Smith & Wesson	.38 Automatic
	.38 Short Colt
	.38 Long Colt
	.38 Special
	.38 Special +P
.38 Special	.380 Automatic
	.357 Magnum
	.38 Special +P
.45 Automatic	.44 S&W Special
	.44 Remington Magnum
.45 Colt	.44 S&W Special
	.44 Remington Magnum
.38-40 Winchester	.38-40 High Velocity
.44-40 Winchester	.44-40 High Velocity

Centerfire Rifles

Chambered for:	*Dangerous ammo. applictions*
.17 Remington	.221 Remington Fireball
	.30 Carbine
.17-223 Remington	.17 Remington
	.221 Remington Fireball
	.30 Carbine
.223 Remington	.222 Remington,
	5.56 military (see Chapter 4)
.243 Winchester	.250-3000 (.250 Savage)
	.225 Winchester (.300 Savage)
6mm Remington (.244 Remington)	.250-3000 (.250 Savage)
.257 Roberts	.250-3000 (.250 Savage)
6.5mm Remington Magnum	.300 Savage
.264 Winchester Magnum	.270 Winchester
	.284 Winchester
	.308 Winchester
	.303 British

<table>
<tr><td></td><td>.350 Remington Magnum
.375 Winchester</td></tr>
<tr><td>.270 Winchester</td><td>.30 Remington
.30-30 Winchester
.300 Savage
.32 Remington
.308 Winchester
7mm Mauser (7×57mm)
.375 Winchester</td></tr>
<tr><td>7mm Mauser (7×57mm)</td><td>.300 Savage</td></tr>
<tr><td>7mm Remington Magnum</td><td>7mm Weatherby Magnum
.270 Winchester
.280 Remington
.35 Remington
.350 Remington Magnum</td></tr>
<tr><td>.280 Remington</td><td>.270 Winchester
.30 Remington
.30-30 Winchester
.300 Savage
.308 Winchester
7mm Mauser (7×57mm)
.375 Winchester</td></tr>
<tr><td>.284 Winchester</td><td>.300 Savage
7mm Mauser (7×57mm)</td></tr>
<tr><td>.30-40 Krag</td><td>.303 Savage
.303 British
.32 Winchester Special</td></tr>
<tr><td>.30-06 Springfield</td><td>8mm Mauser (8×57mm)
.32 Remington
.35 Remington
.375 Winchester</td></tr>
<tr><td>.300 H&H Magnum</td><td>.30-06 Springfield
8mm Mauser (8×57mm)
.30-40 Krag
.375 Winchester
.338 Winchester Magnum</td></tr>
<tr><td>.300 Weatherby Magnum</td><td></td></tr>
<tr><td>.300 Winchester Magnum</td><td>8mm Mauser (Round-nose bullet)
.303 British
.350 Remington Magnum
.38-55 Winchester</td></tr>
<tr><td>.303 British</td><td>.32 Winchester Special</td></tr>
<tr><td>.303 Savage</td><td>.32 Winchester Special
.32-40 Winchester</td></tr>
<tr><td>.308 Winchester</td><td>.300 Savage</td></tr>
<tr><td>.338 Winchester</td><td>.375 Winchester</td></tr>
<tr><td>.348 Winchester</td><td>.35 Remington</td></tr>
<tr><td>.38-55 Winchester</td><td>.375 Winchester</td></tr>
</table>

.375 Winchester   38-55 Winchester
.41 Long Colt

Rimfire Ammunition
Chambered for: *Dangerous ammo. applications*
.22 Win. R.F.   .22 BB Cap
                         .22 CB Cap
                         .22 Short
                         .22 Long
                         .22 Long Rifle
                         .22 Long Rifle Shot
.22 Win. Mag. R.F.   Same as .22 Win. R.F.
.22 Winchester Auto.   Same as .22 Win. R.F.
5mm Remington   Same as .22 Win. R.F. plus
R.F. Magnum   .22 Winchester Auto.
.25 Stevens Long   5mm Remington R.F. Magnum

Shotguns
Chambered for: *Dangerous ammo. applications*
.410 Bore   .219 Zipper
                         .30-30 Winchester
                         .303 British
                         .32 Winchester Special
                         .32-40 Winchester
                         .35 Winchester
                         .38-40 Winchester
                         .44 S&W Special
                         .44-40 Winchester
                         .44 Remington Magnum

Note:

It is, of course, always dangerous to shoot any gauge shotgun shell in any shotgun of a different gauge. Also the use of any shotgun shell longer than the chamber can be dangerous. That is to say, 3-inch shells should never be used in 2¾ chambers, or 2¾-inch shells should never be used in $2\frac{9}{16}$ chambers, and so on.

It is not always possible to emphasize a warning strongly enough.

In reviewing the foregoing cautions I recall an instance when I was a young man. I was driving to the Adirondack Mountains in New York State to collect the winter's venison supply. It was very late, or rather quite early in the morning, when I entered a small town on Route 9 some distance north of Albany. There, alongside the road, was a flashing billboard sign stating "speed limit 35mph—strictly enforced by radar." Well, I wasn't impressed very much and continued along at 60 mph. About 1 mile

PREVENTING SHELL MISMATCHES—Cutaway shotgun shows how, in the heat of hunting excitement, a hunter can borrow a 20-gauge shell from a companion and use it in a 12-gauge gun. When the gun will not fire, he often loads the gun again with a 12-gauge shell and then fires. Resulting pressure can burst the gun barrel.

The wrong cartridge in the wrong chamber can end up looking like this—or worse.

later there was another flashing sign of billboard proportions which simply stated "You have been warned—P.D." Well, I gave that second sign a lot of thought while I continued through the night in a fashion I thought suitable for the circumstance. A mile or so later, and with a newly acquired traffic citation sitting tucked behind the sun visor, I suspect the real meaning of the second sign finally sank into my young but rather thick skull.

The misuse of ammunition can lead to far more costly consequences than the few dollars that traffic citation cost. The author respectfully advises the reader that he/she has passed the second billboard and that "you have been warned!"

CHAPTER ELEVEN

Practical Cartridge and Shell Selections

If advice on the selection of a cartridge for a specific purpose is what you require, or if you are interested in this author's opinions on the only cartridges that fulfill every shooter's need, you have come to the right chapter. I would like to say, however, in defense of my opinions that they are based on 26 years in the firearms industry. They have resulted not only from personal experience in the field and in the ballistic labs but also from literally hundreds of thousands of people with whom I have spoken or corresponded during the last 26 years.

The cartridges and shell applications shown in this chapter shrink the manufacturers' listings considerably. For the sake of brevity, only the best of the lot have been listed, with an explanation to the reader on why the selection was made. Reasons for the omissions would be too lengthy for our purposes here.

RIMFIRE

.22 Long Rifle

This .22 rimfire will do the entire job—target shooting, plinking, and hunting—and it will do it as well as, or better than, any other rimfire when one considers all the possible applications.

CENTERFIRE RIFLE

.22 Hornet
45-Grain Bullet—Varmint

This cartridge is one of the all-time greats with respect to accuracy. It will also get the job done quickly and cleanly to 125 yards. A skilled rifleman can stretch the range another 20 or so yards if he can keep his shots on the head of his quarry.

Almost no recoil, light noise level, very good accuracy, and sufficient velocity make this an excellent choice for any varmint hunter who enjoys stalking his quarry.

Any good big-game cartridge with a heavily constructed bullet will do nicely for most African game—except elephant, Cape buffalo, and rhino.

.222 Remington
50-Grain Bullet—Varmint

For the varmint shooter who can hold his shots to 225 yards (250 yards for the good rifleman), this cartridge is the ideal intermediate-level varmint cartridge. This cartridge also fits into the all-time-great category with respect to accuracy. Noise level is moderate and recoil hardly noticeable.

.22-250 Remington
55-Grain Bullet—Varmint

The .22-250 is the ultimate .22-caliber, long-range varmint round. With excellent accuracy it will get the job done to about 350 to 400 yards. The small amount of recoil and a good muzzle report will not hamper an experienced shooter.

.243 Winchester
80-Grain Bullet—Varmint
100-Grain Bullet—Light Big Game

The .243 is a good accurate round which offers the maximum in range for a varmint gun combined with a good light big-game rifle. This cartridge is an excellent choice for the shooter who considers varmint his primary choice but who wants a more than adequate deer rifle. Of course, antelope or any other light big game can be taken easily with the .243.

.250-3000
87-Grain Bullet—Varmint
100-Grain Bullet—Light Big Game

This cartridge fits the same applications as the .243. However, I favor it slightly over the .243, because barrel life is better due to the lower intensity of the cartridge. It also will, gun for gun, usually outperform the .243 with respect to accuracy.

Elephants have been taken with everything from the 6.5 Mannlicher-Schoenauer to the .600 Nitro Express. However, a .375 H&H or a .458 Winchester Magnum with full metal-cased bullets would perform nicely under the same conditions.

7mm Mauser (7×57mm)
175-Grain Bullet—Light Big Game
The 7mm Mauser is another of the all-time-great cartridges with respect to accuracy. Regrettably, due to its rather modest popularity, the ammo companies don't seem to do their best in loading this round. It must be handloaded if one wants to obtain the super accuracy inherent in this cartridge.

However, even with factory-loaded ammunition, it is a very comfortable rifle to shoot, and accuracy is sufficient to take all light big game at ranges to 250 yards.

.30-30 Winchester
170-Grain Bullet—Light Big Game
This cartridge is the all-time great with respect to popularity. It will always get the job done up to 200 yards on all light big game. In the few bolt-action guns that were built for this cartridge, accuracy was of the tack-driving type. Most of the lever guns available will offer sufficient accuracy to take advantage of the 200-yard practical limit of the cartridge. It is an ideal cartridge for short-range deer hunting, if the hunter will take the time to place his or her shot well.

.308 Winchester
125-Grain Bullet—Varmint
150-Grain Bullet—Light Big Game
180-Grain Bullet—Big Game
To list this cartridge in favor of the .30-06 will create a roar among some shooters. But in a rifle with a 1-12-inch twist it will outshoot the .30-06 on a gun-for-gun basis.

In this sort of situation, nothing less than a .458 would make the hunter feel comfortable.

The 125-grain bullet allows the serious big-game hunter to enjoy off-season practice on varmints with sufficient accuracy to do the job to about 225 yards. On big game it will perform admirably.

The short case allows the hunter the pleasure of a short action and the weight savings that go with such short-action rifles.

Only on our largest bears would I give the .30-06 (220-grain bullet) the edge over the .308. However, for our largest bears neither cartridge is sufficient.

.375 H&H Magnum
270-Grain Bullet—Big Game, Dangerous Game
300-Grain Bullet—Dangerous Game

If you were fortunate enough to hunt big game the world over and wanted to do it with only one rifle, the .375 H&H could go the distance.

Recoil is, of course, brutal; but such is the case with any cartridge capable of taking dangerous game. The 270-grain bullet is to be favored for all applications except where a full metal case bullet is desired, such as for elephant. Then the 300-grain full metal case bullet should be selected.

The .375 H&H is capable of amazing accuracy if the rifleman is up to the recoil and noise. Riflemen who can't handle this kind of recoil should not hunt dangerous game.

.45-70 Government
405-Grain Bullet—Light Big Game
For the serious hunter of light big game, the .45-70 will do the job to about 125 yards. This is an ideal cartridge for the hunter who enjoys the thrill of the stalk and the pleasure of one well-placed shot. It should not be used by a deer shooter who must take every shot that comes along.

.458 Winchester Magnum
500-Grain Bullet—Dangerous Game
510-Grain Bullet—Dangerous Game
For those who hunt elephant, rhino, and Cape buffalo, the .458 is the ultimate cartridge. However, most such hunters will need to do a bit of practice to develop the shooting skill this big gun requires. The recoil, for most shooters, is simply unmanageable.

CENTERFIRE HANDGUNS

9mm Luger
95-Grain Bullet | Varmint, Small Game,
100-Grain Bullet | Self-protection
For those who favor a semiautomatic handgun, the 9mm Luger is an excellent choice. Recoil is modest and accuracy is excellent. For all the purposes mentioned, the 9mm will serve more than adequately.

.357 Magnum
110-Grain Bullet—Varmint, Small Game,
Self-protection
For those who can practice enough to overcome the somewhat annoying muzzle blast and the recoil, the .357 is an ideal handgun cartridge. For the purposes listed, it is perhaps the best choice for the one-handgun shooter—if noise and recoil will not hamper his shooting.

.38 Special
148-Grain Bullet—Target Shooting, Small Game
110-Grain Bullet—Varmint, Small Game, Self-protection
The two loads shown should cover the entire range of applications of a handgun nicely. Its modest recoil and noise level will make it a better choice than the .357 for most shooters. Of course, all .38 Special loads can be fired in the .357 Magnum. Thus, when purchasing a revolver, it is wise to consider the .357 if a possibility exists of some day needing the extra punch of the .357.

SHOTSHELLS
Due to the great overlapping of shotshell loads, few shooters understand or recognize the various load designations such as 2¾–3¾–1¼–2¾ inches. Maximum–1½ inches. In order to help guide you quickly to a good selection, we will list the game to be hunted first, then the correct shot size, followed by the specific loads which will prove satisfactory under all conditions.

Game	Shot Size	Loads
Geese Turkey Fox	BB, 2	• all 12-gauge, 3″ loads. • all 12-gauge, high-velocity loads using 1½ ounces of shot.
Ducks	2, 4, 5	• all 12-gauge high-velocity loads of 1¼ (or more) ounces of shot. • with No. 5 shot the 20-gauge, 3″ magnum with 1¼ ounces of shot will suffice if ranges do not exceed 35 to 40 yards. However, this load will not prove as effective as the high-velocity 12-gauge, 1¼-ounce loads due to the great difference in velocity.
Pheasants	4, 5, 6	• all 12-gauge high-velocity loads of 1¼ (or more) ounces of shot. • with No. 5 or No. 6 shot, the 20-gauge, 3″ magnum with 1¼ ounces of shot can be used.
Squirrels	5, 6	• all 12-, 16-, or 20-gauge loads using 1¼ ounces of shot.
Rabbits	4, 5, 6, 7½	• all 12-, 16-, or 20-gauge loads using 1⅛ to 1¼ ounces of shot. With 7½ size shot all 12-, 16-, or 20-gauge loads using 1 to 1¼ ounces of shot.
Crows Large Grouse	6, 7½	• all 12-, 16-, or 20-gauge loads using 1⅛ ounces (or more) of shot.
Partridge	7½, 8	• all loads using 1 to 1⅛ ounces of shot.

Game	Shot Size	Loads
Doves Pigeons Trap	7½	• all loads using 1⅛ ounces of shot; for trap use either a 12-gauge, 2¾ dram, or 3 dram, 1⅛-ounce load.
Large Rails Woodcock Snipe Quail	8, 9	• all loads using ⅞ to 1⅛ ounces of shot.
Skeet	9	• all standard velocity loads from ⅞ to 1⅛ ounces of shot.
Small Rails (sora)	11	• all loads of standard velocity from ¾ ounce to 1⅛ ounces of shot.

The foregoing listing brings cartridge or shell selection down to an absolute minimum. Can selection be that easy? Yes. Can other cartridges or loads be used? Yes. But this listing is all that is needed to get the whole job done for each area named.

Broadening the selection is more a matter of hobby, gun buggery, or romance. There surely is nothing wrong with being involved in a hobby, being a gun bug, or being in love with a particular loading. However, the preceding list is a practical guide which will ensure success with a minimum of fuss.

Perhaps such a short listing takes some of the enjoyment from the game. If, for example, your favorite light big-game load is a .300 Savage with a 180-grain bullet, I'd be the first to say you are adequately gunned. But for the beginner or expert, the foregoing will always do the job and keep the gun collection down to something less than the price of a new car.

CHAPTER TWELVE

Reloading Ammunition

The handloading of centerfire cartridges can offer many advantages to the shooter. The greatest of these advantages may perhaps be the savings in cost. It is not unreasonable to expect to save at least 50% of the cost of factory ammunition when reloading. In some instances the cost savings can run as high as 75% of the price of a box of factory ammunition.

The money saved by reloading can, of course, be left in your pocket to be enjoyed for other things. However, this is seldom the case. When one begins to reload, the money saved is usually used to purchase more components and hence, enjoy more shooting. This perhaps is the most important benefit of reloading. More shooting will enhance your skills as a marksman with rifle, pistol, or shotgun. In fact, to become a highly skilled shot, it is almost a prerequisite that you handload because of today's cost of ammunition. However, handloading is not for everyone.

Handloading requires meticulous care and attention, if accidents are to be prevented. When properly conducted, handloading is not dangerous. Undertaken with less than the proper outlook and approach, it could easily become a very dangerous pastime.

Handloading has another side to it; it can develop into a hobby. Those who are interested in maximum performance (oftentimes outdoing the accuracy of a factory load) will soon find that their tools and accessories very rapidly become highly sophisticated. A reasonable amount of sophistication is essential to ensure factory-like performance. With each step of upgrading, the hobby interest develops to a greater scope.

The depth of handloading cannot be covered in one chapter; the topic demands a lengthy book. Therefore, it is not the purpose of this chapter to teach you how to become a complete reloader, but rather to point out some of the advantages and disadvantages to handloading. It is also most important to be aware of some basic

Johanna Gordon, an avid benchrest shooter, gets ready to let go with a reload. For benchrest shooting, factory ammunition is unsuitable. Only the very finest of carefully assembled handloads will allow you to be competitive.

cautions that should become a part of your approach to handloading.

For a basic undertanding of how to handload, the author suggests that the reference section of the *Lyman #45 Metallic Reloading Manual* or the reference section of the *Lyman #1 Shot Shell Manual* are a *must* on the reading list.

COMPONENTS

BULLETS

Be sure that the bullets being used are the recommended diameter and weight for the load you are using. Do not trust what the factory box indicates as its contents. Measure a sampling of bullets with a micrometer and also weigh a sampling on your powder scale. Factory errors, while rare, do occur.

Do not interchange brand or style bullets within the same load. Whenever any component change is made, you must return to the suggested starting load and work up, in small increments, to the maximum load for your gun. Never exceed the powder manufacturer's maximum load. Use only the specific bullet listed in your data source.

PRIMERS

Inspect each primer for the presence of anvils before seating. Be sure the anvils are not cocked or tipped.

Primers are potentially dangerous. Be sure to keep only a minimum amount on your workbench. It is wise never to exceed 100 primers. Store primers in the original factory cartons only. Primers stored in bulk in containers, such as glass jars or what have you, amount to no less than bombs on their way to an explosion.

Primers should be kept out of the reach of children and stored in a cool, dry place.

Use only the specific primer listed in the powder manufacturer's data or other reliable data source. Never substitute large-rifle primers for small-rifle primers, etc.

CASES

Do not mix case brands. Ideally, cases should be segregated into lots. When more than 4 or 5% of the cases in a lot show deterioration, the entire lot should be discarded.

One of the most common errors of beginning handloaders is that they are not aware that cases must be kept trimmed to a fairly tight tolerance range.

Experienced reloaders often err by not realizing that there is a limit to the number of times a case can be trimmed. The brass that's being removed by trimming has to come from someplace. That someplace becomes weaker as the brass is removed. In general, four trimmings from the maximum case length to the suggested case length seem to be about right. When the fifth trimming is required, the brass is ready to be discarded.

Cases should be carefully inspected before each loading. The handloader should ascertain that no cracks, splits, stretch marks, or separations are evident. Never load cases that have such damages.

Do not ream or enlarge the primer flash holes.

Shotshells should be carefully inspected for head damage, tube splits, pinholes, and damaged base wads (where appropriate) before reloading.

Never load any case that has a loose primer pocket or one that shows evidence of gas leakage around the primer.

POWDER

Powder should be kept in a cool, dry place and always in the original container.

Use data only from the powder manufacturer or another reliable source.

Never attempt to interpolate data. The use of so-called burning-rate charts for interpolation can be dangerous.

Always start at the suggested starting load and increase your powder charge in small increments until you reach the listed maximum. Never exceed the listed maximum. If your combination of load and firearm shows pressure before you reach the maximum, reduce the charge a full 5%. If this 5% takes you below the suggested starting load, discontinue the use of that load. If trouble persists, look for the cause of the problem. Enlisting the aid of an experienced reloader is always a good double check.

Where your data source does not list a suggested starting load, reduce the maximum charge by 10% for a starting point, EXCEPT where otherwise cautioned by the data source.

Always store powder out of the reach of children and in an area free from potential fire hazards. Never use powder (or any component) whose identity is not certain. Always keep all components in their original factory containers.

Do not purchase powder in paper bags, re-marked cans, etc.

Learn to identify deteriorating powder and discard it promptly and safely.

A booklet on smokeless powder is available from SAAMI, 420 Lexington Avenue, New York, NY 10017. Simply request the smokeless-powder pamphlet and supply a self-addressed stamped envelope. (A similar booklet is also available for primers.)

All powder charges should be checked with a good scale.

WADS

Use only the specific wad listed in your data source. Do not make any substitutions.

SHOT

Shot-charge weight should be carefully checked with a scale. Far too many shotshell reloaders make the very serious error of not owning a good scale.

LOADING OPERATIONS

GENERAL

Avoid any and all distractions, no matter how slight, while reloading. This is not the time to enjoy your favorite music, conversation with a friend, or to have children playing about you.

NEVER smoke in the loading area.

Good housekeeping is a must. Clean up all spilled powder or primers immediately. Primers will dust during handling. Wipe this residue up after each loading session. Use an oily rag to wipe the bench area and tooling.

DECAPPING

Use the proper tools to prevent enlarging of the case's flash hole. Carefully examine flash holes for roundness, burrs, or enlarging before repriming.

RESIZING

Lubricate the case sparingly in order to avoid oil dents.

Check the case's overall length *after* sizing. Trim as necessary.

Do not resize one case from a case of another caliber unless you are totally familiar with and/or equipped to accomplish headspace gauging, neck turning or reaming, case capacity variations, etc.

Proper sizing dies and shell holders eliminate headspacing problems. Nevertheless, tool manufacturers can make mistakes. Be wary.

PRIMING

Never use automatic primer feeds unless you are adequately shielded from them. Then carefully adhere to the manufacturer's recommendation as to the maximum number of primers to be in the unit at any one time.

Before reloading, inspect the primer pockets and clean them where necessary. Be certain never to remove any brass from the case when cleaning primer pockets.

Metallic primers should be seated slowly and uniformly to a depth of .003 inch to .008 inch below flush, in order to properly arm them. NEVER prime cases with a mallet or hammer. This author has interviewed more than one party who was injured when so doing.

Cases should be held by the rim or a vented punch to ensure safety should a primer ignite during seating.

A transparent shield of lucite between the priming station and the operator is good insurance. Safety glasses are a minimum requirement.

If primers do not seat uniformly, locate the problem or replace tooling as required.

POWDER CHARGING

All cases should be carefully inspected for foreign objects before dropping a charge into the case.

If a powder measure is to be used, ten charges should be thrown and weighed as a single unit to determine the nominal charge weight being thrown. Simply divide the total weight by ten to obtain your average charge weight. Thereafter, about every tenth charge should be weighed to ensure you are throwing uniform charges and that the measure has not lost its adjustment.

Uniformity in operation is the secret to success with powder measures.

After charging, each case should be carefully inspected for uniform powder charges. This will prevent overcharging, double charging, and light charging.

It is strongly suggested that a simple powder-height gauge be made. This need be nothing more than a dowel that will fit into the case mouth. A line on the dowel can be marked as a reference guide.

Always refer to the powder manufacturer's data (or another reliable source) to determine the charge weight required. Do not trust your memory as to what powder charge is required.

BULLET SEATING

Seat bullets to recommended loaded lengths only. Excessively long or short seating can cause problems and create safety hazards.

Never use anything but flat or extremely blunt bullets in a tubular magazine. Full metal case bullets can be used in tubular magazines only if they have a flat-nosed profile.

Cartridges to be used in tubular magazines should only be loaded with bullets having a sufficient cannelure to allow proper crimping of the case to the bullet. Smoothsided or shallow cannelured bullets cannot be properly crimped.

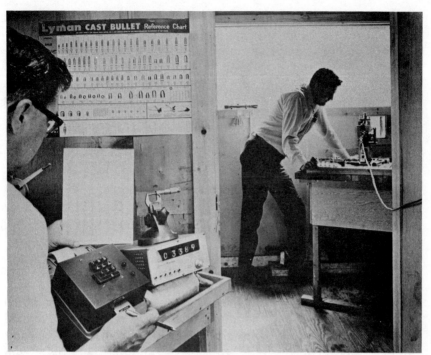

The center of development for the bible of reloaders: The authors check over a pressure testing set-up during the development of the data for the Lyman 45th handbook on reloading.

Be sure bullets are seated firmly to prevent bullet jump, creep, or set back. Make certain seated bullets are not touching the rifling.

HANDLOADING DO'S AND DON'TS

- Do be safe rather than sorry.
- Do throw out anything about which you have any doubts.
- Do use only powder manufacturer's or other reliable sources for your data.
- Don't use data picked up in gun shops, via word of mouth, *or* from magazines, etc.
- Do have your gun carefully examined by qualified personnel if there is any reason to believe you have fired loads which have exceeded the suggested pressure levels.
- Don't use reloads given to you by others. They may not be safe in *your* gun.
- Do examine fired cases after each firing for telltale signs of excessive pressure or brass fatigue.

- Do discontinue the use of any load which performs in an unexpected or unusual manner.
- Don't chamber any round that will not chamber easily.
- Do not attempt to load steel shot until the powder manufacturer's supply data and the necessary components become available. At this writing it is impossible to reload steel shot safely.
- Do keep accurate and detailed records of all loads.
- Don't use any component unless you are 100% sure of its identity.
- Don't use old or deteriorating components.
- Don't use any components that have been stored under adverse conditions such as high temperatures or high humidity.
- Do avoid wildcat cartridges for which no standard specifications are available.

Handloading can be safe, fun, and economical. However, when improperly approached, handloading can be dangerous. Over 3 million people in the United States are reloaders. It doesn't require any special skills to derive all the benefits from handloading. Simply follow the powder manufacturer's recipes, take your time, and be meticulous about details. Under these conditions, handloading can, and will, add a great deal of pleasure to your shooting.

CHAPTER THIRTEEN

New Ammunition Developments

During the past year or so, a number of new ammunition products have been made available to the shooter that have not been covered in earlier portions of this text.

This chapter will attempt to cover, by manufacturer, the major new innovations or modifications in the ammunition industry.

FEDERAL CARTRIDGE CORPORATION

April 27, 1979, marked the 57th anniversary of the founding of the "Ammo Only" company—Federal Cartridge Corporation.

Federal Cartridge Corporation was founded on the remains of an earlier venture. In 1916, the Federal Cartridge and Machinery Company (later Federal Cartridge Company) was established in Anoka, Minnesota. After an ambitious beginning, the company failed, and the facilities sat idle for several years until the reorganization.

By late 1922, the new corporation was selling its "Hi-Power" shotshells. In 1924, rimfire .22's were added to the line. The years since have seen Federal Cartridge grow from a small company into one of the world's major producers of ammunition.

In addition to producing sporting cartridges, Federal Cartridge at Anoka began to manufacture ammunition for the United States government during World War II. However, Federal's major military production has been centered 10 miles east of Anoka at New Brighton, Minnesota.

Then, in 1941, the company contracted to build and operate a large government-owned plant for the manufacture of military small-arms ammunition. This facility, now known as the Twin Cities Army Ammunition Plant, was operated through World War II, with peak employment reaching 25,000 and production reaching 10 million rounds a day. Deactivated in 1945, the plant was reactivated in 1950 for the Korean War and again in 1965 for

Federal .45-70 Government with 300-grain bullet.

the Vietnam War. Cartridges produced include: .50 caliber, .30 caliber, .45 caliber, 7.62mm, and 5.56mm. Through this production, the headstamp of TW became known to hundreds of thousands of American servicemen.

Federal's line of ammunition was completed in 1963. With experience gained from production of military small-arms ammunition, the company introduced its brand of centerfire rifle and pistol cartridges in that year.

.45-70 GOVERNMENT AMMUNITION

The .45-70 Government has been recently added to Federal's line of centerfire ammunition. Adopted by the U.S. Army in 1873, the .45-70 Government cartridge helped tame the West, fought with General George Custer at the Little Big Horn, and served through the Spanish-American War before being replaced by

Federal's newest .44 Remington Magnum cartridge features a 180-grain jacketed bullet.

more modern military calibers. Now well into its second century, the .45-70 Government remains a popular caliber for short-range hunting for light big game.

The .45-70 Government cartridge is normally factory loaded with a 405-grain bullet at a muzzle velocity of about 1330 fps. During the late 1800's, hunters found that bullets weighing about 300 grains offered higher velocity, more energy, and flatter trajectory at close ranges than heavier bullets in .45-70 Government caliber. Federal's load is along these lines. They are using a 300-grain jacketed hollow-point bullet at a muzzle velocity of 1810 fps.

Bullet Wgt. and Style	Velocity in fps			Energy in ft.-lbs.			Bullet Drop	
	Muzzle	100 yds.	200 yds.	Muzzle	100 yds.	200 yds.	100 yds.	200 yds.
300 gr. JHP	1810	1410	1120	2180	1320	840	−6.2"	−22.0"

6mm REMINGTON AMMUNITION

Cartridges in 6mm Remington caliber are also now available from Federal. Two bullet weights are offered: an 80-grain soft point and a 100-grain "Hi-Shok" soft point. The heavier bullet is best for deer and other light big game.

.44 REMINGTON MAGNUM HIGH-VELOCITY AMMUNITION

A new, high-performance .44 Remington Magnum cartridge has also been introduced by Federal. Featuring a 180-grain jacketed hollow-point bullet, muzzle velocity is 1610 fps and muzzle energy 1045 ft.-lbs. from a 4-inch vented test barrel. Obviously, higher velocities may be obtained from handguns with longer barrels.

For hunting varmints and small game, the new Federal load offers high velocity, good retained energy, and flat trajectory. This load is suitable for use in .44 Remington Magnum caliber carbines.

Bullet Wgt. and Style	Velocity in fps		Energy in ft.-lbs.		Midrange Trajectory 50 yds.
	Muzzle	50 yds.	Muzzle	50 yds.	
180 gr. JHP	1610	1365	1045	750	0.5"

.38 SPECIAL H.V. (+P) WITH 158-GRAIN LEAD, SEMI-WADCUTTER BULLET

A .38 Special high-velocity (+P) load with a 158-grain lead, semi-wadcutter bullet has been added to Federal's line of pistol ammunition. Muzzle velocity of the new load is about 915 fps from a 4-inch-length vented test barrel.

PREMIUM RIFLE AMMUNITION

Two years ago Federal introduced a line of premium centerfire rifle cartridges with boattail soft-point bullets (BTSP) in eight calibers. During 1978, three new lighter-weight bullets suitable for hunting varmints and light big game were added. In addition, one new caliber—the .22-250 Remington—was introduced.

The new premium .22-250 Remington cartridge is offered with a 55-grain boattail hollow-point bullet.

In .243 Winchester caliber, an 85-grain boattail hollow-point bullet is now available in addition to the previously offered 100-grain boattail soft point.

A 130-grain boattail soft-point bullet in .270 Winchester caliber is new for 1978, supplementing the 150-grain BTSP in this caliber previously offered in the premium line.

For hunters requiring a high-velocity 7mm Remington Magnum cartridge, Federal now offers a 150-grain boattail soft-point bullet in this caliber. Previously, premium 7mm Remington Mag-

7MM. MAGNUM, 150 GRAINS   .270, 130 GRAINS   .243, 85 GRAINS   .22-250, 55 GRAINS

Federal Premium centerfire ammunition.

The components of Federal's new Premium shotshell.

num ammunition was available only with a 175-grain BTSP bullet.

All premium cartridges use boattail bullets made for Federal by the Sierra Bullet Company. The streamlined, tapered base of the boattail bullet greatly reduces drag at speeds below the

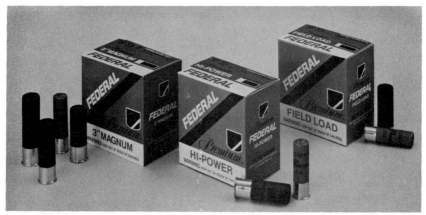

Federal's Premium loads are available in a wide variety.

velocity of sound (1080 fps). This results in higher retained velocity, more striking energy, flatter trajectory, and less wind drift than with conventional flat-base bullets when ranges are extreme.

PREMIUM SHOTSHELL AMMUNITION

Federal's line of premium shotshell ammunition was expanded by the addition of two new 12-gauge loads during 1978. In addition, the 10-gauge premium shell has been modified by increasing the shot charge.

New in 3-inch, 12-gauge magnum is a 1⅝-ounce offering in shot size 4 or 6. Supplementing the previous 3-inch, 1⅞-ounce premium load, the new 1⅝-ounce shell has all the premium magnum features designed to provide serious hunters with the highest performance. These features include copperplated, extra-hard lead shot, granulated plastic buffer material in the shot charge, and Federal's Magnum Triple-Plus wad column.

The 10-gauge, 3½-inch premium magnum, formerly a 2-ounce load, is now offered with 2¼ ounces of shot at the same muzzle velocity of 1210 fps. Shot sizes will be BB, 2, and 4.

For upland birds and other small game, Federal has introduced a new premium 12-gauge 3¼-dram equivalent, 1¼-ounce heavy field load utilizing copperplated, extra-hard shot and a triple-plus plastic wad column, but without buffer material in the shot charge. This heavy field load fills a gap between two other premium loads in 12-gauge: the Hi-Power 3¾ dram equivalent,

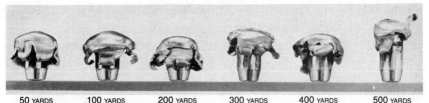

| 50 YARDS | 100 YARDS | 200 YARDS | 300 YARDS | 400 YARDS | 500 YARDS |

The photographs show the expansion of Sierra boattail bullets used in Federal's Premium centerfire cartridges.

1¼ ounce; and the light field load 3¼ dram equivalent, 1⅛ ounce.

REMINGTON ARMS COMPANY, INC.

The year 1816 marked the founding of Remington Arms Company, Inc., the oldest gunmaker in America. From its beginning in a small forge near what is now Ilion, New York, the concern, founded by Eliphalet Remington, has grown to its present position as one of the world's leading producers of sporting firearms and ammunition.

Eliphalet Remington's skill as a gunsmith led him into the business at a time when there were almost no firearms made in America.

In 1888, Hartley and Graham, a major distributor of sporting goods, purchased a large interest in the Remington firm. A year later, following the death of Philo Remington, the founder's son, complete control of the company came into the hands of Marcellus Hartley. In addition to his interest in Hartley and Graham and Remington, Hartley was the founder and owner of the Union Metallic Cartridge Company of Bridgeport, Connecticut. In the days immediately following the Civil War, this company was organized to manufacture metallic cartridges for the newly invented breech-loading rifles. Remington pioneered the development of breechloaders, and it was thus logical that the two companies should come under common ownership.

Although both organizations were owned by the Hartley interests, they were not physically merged until 1912. In 1920, the concern, previously known as Remington Arms-Union Metallic Cartridge Company, was reorganized as Remington Arms Company, Inc.

In 1933, E. I. du Pont de Nemours and Company, long a supplier of sporting powder for ammunition, purchased a controlling interest in Remington. In the same year, Remington acquired the business of Chamberlin Cartridge and Target Company, manufacturers of clay targets and traps used to propel them. The Peters Cartridge Company, now operated as the Peters Cartridge Division of Remington Arms Company, Inc, was acquired in 1934.

In addition to firearms, ammunition, traps, and targets, Remington manufactures "Hi-Dense" powdered metal parts and tungsten-carbide-coated abrasive cutting products. It has a wholly owned subsidiary, Remington Arms of Canada Limited, which produces sporting ammunition and distributes other company products. It has a 50 percent interest in Companhia Brasileira de Cartuchos, a Brazilian producer of firearms, ammunition, and metal articles. It also has a 40 percent interest in Cartuchos Deportivos de Mexico, S.A., which produces shotshells, rimfire and centerfire cartridges, lead shot, industrial shells, and a broad range of miscellaneous metal articles for industrial purposes.

The company's corporate headquarters are in Bridgeport, Connecticut. It manufactures ammunition and abrasive products in Bridgeport and ammunition in Lonoke, Arkansas. Its sporting firearms, powdered metal parts, and traps are made in Ilion, New York, while clay targets are produced in plants at Findlay, Ohio; Ada, Oklahoma; and Athens, Georgia. The Lake City Army Ammunition Plant in Independence, Missouri, a government-owned facility that produces military small-arms ammunition, is operated by Remington on a fixed-fee contract basis.

Although Remington has been in business since 1816, it is a vital and vibrant company. Today it continues a long tradition of providing well-designed, carefully made products to meet the needs of hunters and shooters.

BUFFERED SHOTSHELL LOADS

Buffered shotshells produce a tighter pattern than standard shells through the use of a granulated plastic mixed with the shot. This development was particularly important with the larger shot sizes where prevention of pellet deformation is critical because of the lesser number of total pellets in the charge. Winchester-Western was the first to market such ammunition.

Now the same basic kind of pattern-tightening filler is available in Remington's new "Nitro-Mag" 12-gauge, 2¾-inch and 3-inch magnum shells.

The 2¾-inch "Nitro-Mag" shells are loaded with 1½ ounces of

Nitro Mag is Remington's answer to Winchester's Double X loads.

No. 2 and No. 4 pellet sizes. In 3-inch magnum lengths, "Nitro-Mag" shells will house a payload of 1⅝ ounces of No. 2's and No. 4's and a 1⅞-ounce charge of BB's No. 2's, and No. 4's.

Remington claims that "Nitro-Mag" shells will put a minimum of 80% of the pellets in a 30-inch circle at 40 yards from the majority of full-choked, 12-gauge guns.

SPECIFICATIONS FOR REMINGTON "NITRO-MAG" 12-GAUGE SHOTGUN SHELLS

Gauge	Length	Shot Charge	Shot Sizes
12	2¾"	1½ ozs.	2, 4
12	3"	1⅝ ozs.	2, 4
12	3"	1⅞ ozs.	BB, 2, 4

.380 AUTO

In a continuation of its policy of adding greater versatility to the performance of handgun calibers, Remington Arms Company, Inc., has developed a new bullet for the .380 Auto Pistol cartridge.

The new loading consists of an 88-grain jacketed hollow-point (JHP) bullet. With improved expansion characteristics and somewhat increased velocity and energy performance, the .380 Auto 88-grain JHP ammunition now extends the performance potential of this cartridge.

The picture shows Peters' new Blue Magic Target load.

Remington's new 380 auto hollow-point round.

BALLISTICS FOR THE .380 AUTO PISTOL
88-GRAIN JACKETED HOLLOW POINT*

Range (Yds.)	Velocity (fps)	Energy (ft.-lbs.)	Midrange Trajectory
0	990	191	
50	920	165	1.2″
100	863	146	5.1″

*From 3¾″ barrel

Remington's Accelerator (left) alongside a standard .30-06 Spring-field.

NEW REMINGTON "ACCELERATOR"
CARTRIDGE FOR .30-06 RIFLES

A new and unique development by Remington Arms Company, Inc., called the "Accelerator" cartridge, converts a .30-06-caliber rifle to a .22-caliber varmint gun with the highest advertised bullet velocity ever produced in a factory-loaded round in a 24-inch barrel.

The "Accelerator" cartridge involves the use of a .224-inch, 55-grain pointed soft-point bullet inside a .30-caliber sabot casing. The sabot-encased .22-caliber bullet is loaded in a standard .30-06 case and fired in a regular .30-06 rifle.

The muzzle velocity from a 24-inch barrel is 4080 fps, superior to that of the original .220 Swift in the same barrel length. It surpasses, by 350 fps, the velocity of any current factory-loaded .22-caliber cartridge with a 55-grain bullet. The muzzle energy of the "Accelerator" cartridge is 2033 ft.-lbs., the highest ever developed for a .22-caliber bullet.

Once the "Accelerator" cartridge's sabot-encased bullet leaves the muzzle, air resistance peels open the sabot, causing it to drop off the bullet without, according to Remington, affecting its accuracy or trajectory.

The "Accelerator" enables the owner of a manually operated (bolt or pump action) .30-06 rifle to convert that rifle to a varmint gun by simply switching ammunition. "Accelerator" cartridges do not operate the action of autoloading rifles but will function in them on a single-shot basis.

COMPARATIVE CENTERFIRE PERFORMANCE DATA
.30-06 SPRINGFIELD "ACCELERATOR"
VS. CONVENTIONAL AMMUNITION

	"Accelerator" .30-06 55-gr.	.220 Swift 48-gr.	.22-250 Rem. 55-gr.	6mm Rem. 80-gr.	.30-06 Springfield 150-gr.
Velocity fps					
Muzzle	4080	4040	3730	3470	2910
100 yds.	3485	3418	3180	3064	2617
200 yds.	2965	2877	2695	2694	2342
300 yds.	2502	2397	2257	2352	2083
400 yds.	2083	1965	1863	2036	1843
500 yds.	1709	1587	1519	1747	1622
Energy ft.-lbs.					
Muzzle	2033	1739	1699	2139	2820
100 yds.	1483	1245	1235	1667	2281
200 yds.	1074	882	887	1289	1827
300 yds.	764	612	622	982	1445
400 yds.	530	411	424	736	1131
500 yds.	356	268	282	542	876
Bullet Drop Inches					
Muzzle	0.00	0.00	0.00	0.00	0.00
100 yds.	1.16	1.19	1.39	1.57	2.20
200 yds.	5.21	5.39	6.25	6.86	9.51
300 yds.	13.25	13.84	15.96	16.98	23.18
400 yds.	26.92	28.42	32.57	33.43	44.90
500 yds.	48.71	52.16	59.36	58.33	76.93

All data based on 24″ test barrel results.

8mm REMINGTON MAGNUM

Remington now offers hunters a high-powered cartridge with a combination of bullet weight and energy capable of handling all North American big game and most African big game as well (except elephant, buffalo, and rhino).

The new cartridge is the 8mm Remington Magnum, loaded with either 185-grain or 220-grain pointed soft-point "Core-Lokt" bullets. Actual bullet diameter of the new round is .323 inch. Both loadings have sustained downrange velocities that produce exceptionally flat trajectories and high remaining energies for big-game bullets of these weights.

The 220-grain loading of the 8mm Remington Magnum has greater downrange energy (200 yards or more) than 200-grain bullets in either the .300 Winchester Magnum or the .338 Winchester Magnum.

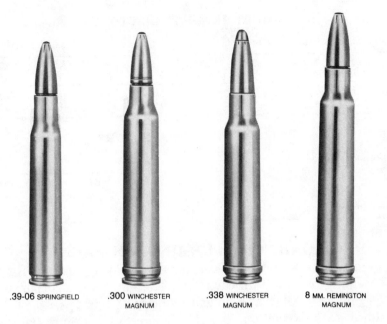

| .39-06 SPRINGFIELD | .300 WINCHESTER MAGNUM | .338 WINCHESTER MAGNUM | 8 MM. REMINGTON MAGNUM |

Muzzle velocity of the 8mm Remington Magnum's 185-grain load, at 3080 fps, combines optimum characteristics of flat shooting and high energy without developing *excessively* uncomfortable recoil in a rifle.

The trajectory of the 8mm Remington Magnum 185-grain load is flatter even than that of a .30-06 with the 150-grain bullet. In comparison, the trajectory of the 8mm Remington Magnum 220-grain load is flatter than the .300 Winchester Magnum 220-grain load and roughly equal to the .338 Winchester Magnum 200-grain load despite the latter's lighter bullet.

With its dual characteristics of flat trajectory and high energy, the new 8mm Remington Magnum provides the hunter with an all-range gun for hunting almost all big-game animals.

BALLISTICS OF 8mm REMINGTON MAGNUM
185-GRAIN AND 220-GRAIN POINTED SOFT-POINT
"CORE-LOKT" BULLETS

Range (Yds.)	Velocity 185 grain	fps 220 grain	Energy 185 grain	ft.-lbs. 220 grain
0	3080	2830	3896	3912
100	2761	2581	3132	3255
200	2464	2346	2494	2688
300	2186	2123	1963	2201
400	1926	1913	1524	1787
500	1687	1716	1170	1439

SHORT-RANGE TRAJECTORY
(SIGHTED-IN AT 150 YARDS)

8mm Rem. Mag. Bullet Weight	50 Yds.	100 Yds.	150 Yds.	200 Yds.	250 Yds.	300 Yds.
185 Grain	+0.5"	+0.8"	0.0"	−2.1"	−5.6"	−10.7"
220 Grain	+0.6"	+1.0"	0.0"	−2.4"	−6.4"	−12.1"

LONG-RANGE TRAJECTORY
(SIGHTED-IN AT 200 YARDS)

	100	150	200	250	300	400	500
185 Grain	+1.8"	+1.6"	0.0	−3.0"	−7.6"	−22.5"	−46.8"
220 Grain	+2.2"	+1.8"	0.0	−3.4"	−8.5"	−24.7"	−50.5"

MEDIUM LOAD FOR .44 REMINGTON MAGNUM

Remington has introduced a medium-velocity loading for the .44 Remington Magnum cartridge. The new load incorporates a 240-grain lead bullet at a muzzle velocity of 1000 fps.

Since its introduction, the .44 Remington Magnum has been the most powerful handgun cartridge commercially available. As such, it provided exceptionally high delivered energy where this was desirable or useful. The heavy recoil makes the cartridge difficult to handle for most shooters.

The new loading, at a medium-velocity range, develops a nominal muzzle energy of 530 ft.-lbs. Adequate for small-game hunting with a more comfortable recoil level, it is also useful for both plinking and target practice. It is just slightly higher in muzzle energy than the .41 Remington Magnum with the 210-grain lead bullet and still greater in total energy than any .38 Special loading.

BALLISTICS FOR .44 REMINGTON MAGNUM
240-GRAIN MEDIUM LOAD

	Velocity fps	Energy ft.-lbs.	Midrange Trajectory (Inches)
Muzzle	1000	533	—
50 yds.	947	477	1.1"
100 yds.	902	433	4.8"

NOTE: Figures produced in 6¼" barrel.

WINCHESTER-WESTERN

Winchester-Western, now part of the Winchester Group of the Olin Corporation, has produced literally billions of rounds of ammunition in its history-making 112 years.

The worldwide arms and ammunition operations of the present Winchester-Western Division include: headquarters and arms manufacturing facilities in New Haven, Connecticut; the ammunition plant (complemented by Olin's brass mill) in East Alton, Illinois; Winchester-Western (Canada) Ltd., producers of Cooey and Winchester Firearms and Western Ammunition, in Cobourg, Ontario; a Winchester shotshell plant in Anagni, Italy; and another ammunition plant in Geelong, Australia. Oli-Kodensha at Tochigi, Japan, produces the Winchester Model 101 over-and-under shotgun and the Model 23 side-by-side shotgun.

Winchester pioneered the most important sporting-ammunition development of the 1800's—smokeless-powder cartridges. When the company brought out its famous Model 1894, it introduced the first commercial smokeless-powder sporting cartridges.

The following year Winchester introduced the Model 1895 lever-action rifle which handled the most powerful of that day's smokeless-powder cartridges.

Wide diversification after WW I, in an attempt to utilize the expanded wartime plant for peacetime purposes, led to a serious drain of the company's resources. The Great Depression of 1929 struck the final financial blow. The Winchester Repeating Arms Company went into receivership on January 22, 1931, and was purchased by the Western Cartridge Company by the end of the year.

With the merger, Winchester-Western became the largest owner of patents on firearms and ammunition developments in the industry. John M. Olin, who had inspired and managed the purchase of Winchester and who was also the first vice president of the Western Cartridge Company, moved to New Haven for four years to personally supervise the revitalization of Winchester.

An innovator as well as an executive, John Olin took an active interest in the development of new arms and ammunition products. By 1940, the company was in a solid position to serve the country once more as World War II approached. By war's end, Winchester-Western produced 15 billion rounds of ammunition.

Today, Winchester-Western remains one of the leaders in the world's sporting firearms and ammunition industry.

.375 WINCHESTER

The new .375 Winchester cartridge will probably become another long-term favorite cartridge. The .375 Winchester car-

The new Winchester 375 (left) alongside the old standby .30-30.

tridge is offered in two bullet weights: a 200-grain power point and a 250-grain power point. According to Winchester, both bullets deliver abut 10 percent more energy on target than the popular .30-30 Winchester or the .35 Remington cartridges. For medium ranges and light big game, this cartridge should prove to be an excellent choice.

The following .375 Winchester ballistics tables were supplied by Winchester-Western.

EXTERIOR BALLISTICS TABLES
.375 WINCHESTER

200-GRAIN POWER POINT BULLET

Range, Yards		0	50	100	150	200	250	300
Velocity, fps		2200	2016	1841	1677	1526	1389	1268
Energy, ft.-lbs.		2150	1805	1506	1249	1034	857	714
Line of sight*								
If sighted-in at	50 yds.	—	0	−1.2″	−5.1″	−11.9″	−22.6″	−37.5″
	100 yds.	—	+0.6″	0	−3.2″	−9.5″	−19.5″	−33.8″
	150 yds.	—	+1.7″	+2.1″	0	−5.2″	−14.1″	−27.4″
	200 yds.	—	+3.0″	+4.7″	+3.9″	0	—	—

250-GRAIN POWER POINT BULLET

Range, Yards		0	50	100	150	200	250	300
Velocity, fps		1900	1770	1647	1531	1424	1326	1239
Energy, ft.-lbs.		2005	1740	1506	1302	1126	976	852
Line of sight*								
If sighted-in at	50 yds.	—	0	−1.9″	−6.9″	−15.7″	−28.7″	−46.5″
	100 yds.	—	+0.9″	0	−4.1″	−12.0″	−24.0″	−40.9″
	150 yds.	—	+2.3″	+2.7″	0	−6.5″	−17.2″	−32.7″
	200 yds.	—	+3.9″	+6.0″	+4.9″	0	—	—

NOTE: Ballistics established with 24-inch barrel.

*With sights 0.9 inches above line of bore.

Winchester's new .375-caliber rifle has a 20-inch barrel. The foregoing ballistics were based on a 24-inch barrel. Mr. Ed. Harris of the National Rifle Association passed the following data on to the author. This data was obtained at 15 feet from the muzzle in an actual 20-inch barreled 94 Big Bore.

200-Grain Bullet

Average Velocity	2115 fps
Highest recorded velocity in 10 shot string	2139 fps
Lowest recorded velocity in 10 shot string	2083 fps
Extreme variation in 10 shot string	56 fps
Standard deviation	19 fps

250-Grain Bullet

Average Velocity	1858 fps
Highest recorded velocity in 10 shot string	1887 fps
Lowest recorded velocity in 10 shot string	1840 fps
Extreme variation in 10 shot string	47 fps
Standard deviation	16 fps

From the above average velocities, the 200-grain bullet has a 15 foot energy of 1987 ft.-lbs., and the 250-grain bullet has a 15 foot energy of 1917 ft.-lbs.

.22-CALIBER XPEDITER

A new higher velocity Super-X Xpediter .22-caliber cartridge has been added to the Winchester-Western ammunition line.

Said to fill the void between the standard high-velocity .22 Long Rifle and the .22 Winchester Magnum Rimfire cartridges, the Xpediter provides the hunter with another small-game and varmint-shooting cartridge for ranges up to 50 yards. It has a 29-grain lubaloy bullet with a hollow-point cavity designed to deliver good energy release. Despite its high velocity, the Xpediter car-

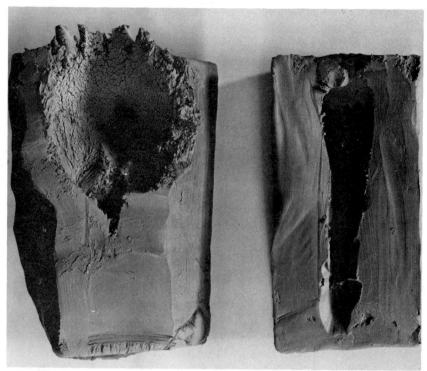

The photograph at left shows the expansion of Winchester-Western's new .22 Xpeditor. At right, the expansion of a standard .22 Long Rifle.

tridge is loaded to pressure limits standardized by the industry for .22-caliber Long Rifle ammunition.

Muzzle velocity of the Xpediter is approximately 1680 fps fired from a 24-inch barrel. Muzzle energy is about 182 ft.-lbs. By comparison, a standard high-velocity Super-X .22-caliber Long Rifle hollow-point cartridge achieves a muzzle velocity of approximately 1280 fps, and delivers a muzzle energy of about 138 ft.-lbs. fired from the same length barrel. The midrange trajectory of the Xpediter is 2.5 inches at 100 yards, an inch less than that of the standard high-velocity Super-X .22-caliber Long Rifle hollow-point cartridge.

The nickel-plated Xpediter cartridge case is somewhat longer than the standard .22-caliber Long Rifle case. *Xpediter should be used only in rifles and handguns with .22-caliber Long Rifle sporting chambers.*

Winchester 10-gauge 3½" Magnum shotshell with 2½ ounces of shot, loaded in the Double X style.

10-GAUGE SUPER-X DOUBLE-X

The 10-gauge XX shell is a 3½-inch poly-formed plastic shell loaded with 2¼ ounces of No. 2 or No. 4 lead shot. Like its 12-gauge predecessors, the new 10-gauge Super-X Double-X Magnum has the Winchester-Western combination of the Mark 5 plastic collar and granulated polyethylene powder. These two features provide the pellet protection that results in denser, more effective patterns at greater distances.

Testing of the 2¼-ounce, 10-gauge load by Winchester-Western ballisticians is said to have produced patterns of more than 80 percent at 40 yards. The load's pellet energy at from 40 to 60 yards averaged approximately 40 percent greater than that of a conventional 10-gauge shotshell without the Super-X Double-X features.

NEW HANDGUN CARTRIDGES

At this writing, Winchester is developing two new pistol cartridges. They are the 9mm Winchester Magnum and a .45 Winchester Magnum. While designed for a gas-operated, semiautomatic handgun, they will probably first be used in the Contender single-shot pistol.

CHAPTER FOURTEEN

Properties of Sporting Ammunition and Recommendations for its Storage and Handling, as Supplied by SAAMI

These pages have been prepared by the Sporting Arms and Ammunition Manufacturers' Institute, based upon information currently available to it. They are furnished to interested persons as a courtesy and in the interests of safety. They are not intended to be comprehensive; they do not modify or replace safety suggestions, standards, or regulations made by designated authorities, public or private. They are subject to revision as additional knowledge and experience are gained. SAAMI expressly disclaims any warranty, obligation, or liability whatsoever in connection with the information contained herein or its use.

These paragraphs are meant to give everyone concerned with the shipment, storage, and handling of sporting ammunition certain basic and important facts about the properties of this widely distributed product. Such information should dispel some of the rumors and tales which persist regarding ammunition bulk safety. It also outlines recommended storage conditions, and reports observations of the reactions of sporting ammunition when exposed to fire or intense heat and rough or vigorous handling.

These statements and recommendations do not supersede local, state, or federal regulations. Local authorities should be consulted regarding any regulations on the storage, transportation, sale, and handling of sporting ammunition in each specific community.

PROPERTIES OF SPORTING AMMUNITION

All sporting ammunition is carefully engineered and manufactured as an article of commerce. It has a specific use; if stored in a proper manner and used as intended in sporting firearms in good condition and adapted for the specific cartridge, the safety and satisfaction of the shooter should be assured.

Sporting ammunition is packed in cartons and cases as specified by the U.S. Dept. of Transportation. Containers of these designs were developed in the interest of safety in transportation, storage, and marketing. Therefore, unapproved packaging should never be substituted.

Specific properties or characteristics of sporting ammunition of particular interest to shippers, warehousemen, dealers, and users are as follows:

*Stocks of sporting ammunition will NOT mass explode. That is to say, if one cartridge or shotshell in a carton or case is caused to fire, it will not cause other adjacent cartridges or shotshells or their packages to explode sympathetically or in a simultaneous manner. There are no limits imposed on packaged quantities of ammunition which may be shipped, warehoused or displayed in commercial establishments. This fact recognizes the inherently safe, nonhazardous characteristics of such ammunition in public or private storage.

*Sporting-arms ammunition is not a super-sensitive item. Packages of ammunition may be dropped from any height which the packages will physically withstand, and cartridges or shotshells therein will not fire due to shock. Properly packaged sporting ammunition will withstand all the rough handling tests of commerce such as drop test, vibration tests, and rotating drum tests without individual cartridges or shotshells firing.

*Modern sporting ammunition, if discharged in the open without the support provided by a firearm's chamber or other close confinement, discharges inefficiently. The flight—more accurately "movement"—of projectiles or debris particles from such incidents are extremely limited in velocity and range. The small primer cups or rimfire-case fragments present the missiles of highest velocity in such occurrences. Specifically, bullets and shot charges, being heavier than shell or cartridge cases in most instances, are rarely projected away from the location at which the unchambered round of ammunition was caused to ignite and discharge. However, small particles of metal or plastic from the burst case and primer cups may be propelled for short distances (usually not over 50 feet) at velocities sufficient in some instances to cause injury or discomfort.

Insofar as the Sporting Arms and Ammunition Manufacturers' Institute has been able to determine, there have been no substantiated reports of serious or fatal injuries caused by the discharge of packaged or loose sporting ammunition in fires, regardless of the quantity or type of cartridges or shotshells involved. SAAMI has no verified report of any fire fighter hurt by flying bullets or shot pellets in fires involving a sportsman's in-the-home personal supply of ammunition, a retail sporting goods store's stock, wholesaler's or distributor's sizable inventory or an in-transit cargo of this product, as long as he was wearing normal fire-fighting protective gear.

HANDLING AND STORAGE OF AMMUNITION

Sporting ammunition contains explosive ingredients: a percussion-sensitive primer mixture and a smokeless propellant. It should be treated with respect and common care in all handling, transportation, and storage.

Ammunition should be stored in the factory carton or package. The labeling and identification on the original container help to assure that future use will be in the gun for which the ammunition is intended.

Sporting ammunition stored in the home, retail outlet, or distributor's warehouse over extended periods in factory packaging and subject to the ordinary variations of temperature and humidity ranging from tropic to arctic conditions can be expected to perform satisfactorily and safely in the firearms for which it was intended, if such firearms are in proper working order and condition. Extremely high temperatures, however, should be avoided.

Ammunition should not be immersed in water or exposed to any organic solvent, paint thinner, petroleum product, ammonia, etc. Such materials may penetrate a loaded round and reach the powder or primer; a deteriorating effect will result which may cause misfires or squib shots. The latter can result in a projectile's lodging in a gun barrel, the obstruction possibly causing serious damage or injury when another shot is fired.

Ideally, home storage of sporting ammunition will be in a locked closet or cabinet out of the reach of children and uninformed or incompetent persons. Both guns and ammunition should always be stored out of sight and the reach of children and all persons not physically or mentally capable to give them correct, proper use and respect.

Animals may be fatally poisoned by chewing on loaded

shotshells, since an ingredient in *some* smokeless powder is toxic to them. Don't discard ammunition in the field.

During short periods when moving to and from the hunting field or target range, storing guns and ammunition in locked auto trunks may be convenient, or required by state or local law. The possibilities of extremely high temperatures make it sensible to remove firearms and ammunition from vehicles at the destination of the trip. The passenger compartment of a closed car when exposed to the sun often develops an extremely high temperature and therefore, is not a desirable spot to leave ammunition.

While blank cartridges will not mass detonate if one in a box is caused to fire, the noise of the firing outside a gun will be nearly as loud as in normal use and may be harmful to hearing. The blank's "explosion" may also be rather violent. Obviously, blank cartridges deserve the same respectful handling and careful storage accorded other ammunition.

Retail and wholesale stocks of ammunition not required for display should be stored in original outer cartons or boxes exactly as supplied by the factory. When placed on a basement or warehouse floor subject to moisture, it would be well to stack the cartons on pallets. Do not stack them more than 3 or 4 feet high and make sure that air circulation is provided. In some locations police or public-security regulations may prescribe the manner in which sporting ammunition stocks are displayed and the quantity that may be in sight. Check with local authorities. Packages of ammunition should not be placed in proximity to heavily trafficked aisles in the reach of children.

SPORTING AMMUNITION IN A FIRE

Although much has been written and rumored about the 4th of July and the so-called havoc of ammunition in fires, it just isn't so. Members of fire-fighting units are understandably uneasy when confronted by fires where ammunition is involved.

Several members of the Sporting Arms and Ammunition Manufacturers' Institute have undertaken extensive experiments to show what can be expected when ammunition is involved in a fire. After such fires, these companies have also made careful investigations which show that the missiles do not have sufficient velocity to penetrate the garments and protective gear worn by fire fighters.

Tests also show that the whizzing sounds heard in the vicinity of ammunition fires are caused by primers expelled from the burning cartridges. The "pops and bangs" are exploding primers; the propellant powders burn inefficiently and make little noise.

Once the packing materials have been consumed, metallic cartridges in a fire are difficult to sustain in a burning condition due to the cooling effects of the metal parts and the relatively high ratio of metal weight to smokeless powder. Only a vigorous fire around metallic ammunition stocks will cause all cartridges to burn. Shotshell ammunition is difficult to ignite, but once well-ignited it will sustain its own burning due to the plastic or paper tubes.

Ammunition that has been in a structural fire and wetted or scorched should never be returned to commercial sales channels or sold at fire sales. It should be scrapped. Never dispose of ammunition by burying or dumping it in a waterway. It may be retrieved years later, fully "live," and pose dangers to children or uninformed persons.

Unserviceable ammunition should be scrapped by incineration in compliance with all federal, state, and local regulations. Closely inspect the remains to ensure destruction of all ammunition.

KNOW THE FOLLOWING RECOMMENDATIONS ON STORAGE AND HANDLING ISSUED BY THE NATIONAL FIRE PROTECTION ASSOCIATION, 470 ATLANTIC AVE., BOSTON, MA 02210 AND REPRINTED WITH THEIR PERMISSION:

CODE FOR THE MANUFACTURE, TRANSPORTATION, STORAGE, AND USE OF EXPLOSIVE MATERIALS NFPA No. 495

CHAPTER 9. Small-Arms Ammunition, Small-Arms Primers, Smokeless Propellants and Black-Powder Propellants

91. GENERAL PROVISIONS

911. In addition to all other applicable requirements in this Code, the intrastate transportation of small-arms ammunition, small-arms ammunition primers, smokeless propellants, and black-powder propellants shall be in accordance with current U.S. Department of Transportation regulations.

912. The provisions of this chapter apply to the channels of distribution of and to the users of small-arms ammunition, small-arms primers, smokeless propellants, and black-powder propellants. They do not apply to in-process storage and intraplant transportation during manufacture.

913. This chapter covers transportation and storage of small-arms ammunition and components. It is not intended to cover safety procedures in the use of small-arms ammunition and components.

92. SMALL-ARMS AMMUNITION

921. No restrictions are imposed on truck or rail transportation of small-arms ammunition other than those which are imposed by the U.S. Department of Transportation or by the presence of other hazardous material.

922. No quantity limitations shall be imposed on storage of small-arms ammunition in warehouses, retail stores, and other general occupancies, except those imposed by limitations of storage facilities and consistency with public safety.

923. Small-arms ammunition shall be separated from flammable liquids, flammable solids (as classified by the U.S. Department of Transportation), and oxidizing materials by a fire-resistive wall of one-hour rating or by a distance of 25 feet.

924. Small-arms ammunition shall not be stored together with Class A or Class B explosives (as defined by U.S. Department of Transportation regulations) unless the storage facility is adequate for this latter storage.

CHAPTER FIFTEEN

Recommendations for Storage and Control of Small-Arms Ammunition for Security and Law-Enforcement Use, as Supplied by SAAMI

These pages have been prepared by the Sporting Arms and Ammunition Manufacturers' Institute, Inc.

There is no maximum storage period for ammunition which has been commercially manufactured in the United States since World War II. Under the *ideal* conditions of a cool, dry atmosphere, free of any corrosive fumes, there should be no deterioration of modern ammunition. Many instances have been recorded where factory-loaded cartridges over 50 years old have performed in conformance with original specifications of the manufacturer. Five-year storage without detrimental effect is well-documented and not uncommon. In other words, age in itself is not a factor. It is the storage conditions that are important. These, even under the best control, are not ideal because they will vary from time to time. For this reason, any ammunition suspected of having been in a long-term storage situation should be examined by an expert before it is fired. If there is any doubt about its condition, do not use it.

Even though modern ammunition is quite stable when stored for extended periods, the storage area temperature and humidity should be moderate and wide variations avoided. Above all, the storage area should be dry. It has been found that one of the best places in your home to store shells and cartridges is on a shelf in your bedroom closet. The poorest place to keep ammunition is on the rear window shelf of an automobile where the sun can shine directly on it, or in the trunk or glove compartment of a car, especially during hot weather.

Very high temperatures and wide temperature variations can cause permanent, internal deterioration which may cause high pressures or other changes in anticipated performance. This alteration is permanent and cannot be reversed even if the ammunition were held at a constant moderate temperature for several hours.

Storage at low temperature will usually not affect the performance *IF* the ammunition is brought up to a reasonable temperature for a few hours before use. Ammunition fired at temperatures below 0° F. (–18° C.) or above 120° F. (49° C.) may suffer some loss in performance.

Ammunition should never be stored in a corrosive atmosphere, such as in or near chemical plants, stables, the seaside, etc.

RIMFIRE

1. GENERAL
When stored for periods of ten or more years, modern rimfire ammunition is quite stable under conditions of moderate temperatures and humidity such as are found in the United States.

High temperatures coupled with high humidities over long periods tend to cause some loss in velocity. This result is permanent and cannot be reversed even if the ammunition were held at a constant room temperature for several hours. Long storage at low temperatures will not usually alter anticipated performance *IF* the ammunition is permitted to reach more reasonable temperatures before use. Ammunition fired at temperatures below 0° F. (–18° C.) or above 120° F. (49° C.) may suffer some loss in performance.

Ammunition should never be stored in a corrosive atmosphere, such as in or near chemical plants, stables, the seaside, etc.

2. CONTROL OF AMMUNITION CARRIED IN PISTOLS OR REVOLVERS IN HOLSTERS
Rimfire ammunition is commonly used for marksmanship practice and is not usually carried on the person for a law-enforcement purpose.

Should rimfire ammunition be carried in pistols and revolvers on the person, it is recommended that the chambers and magazines be dry to minimize possible contamination of the ammunition by oil or solvents (or both).

At least once a week, preferably daily, and certainly after a firearm is used, all ammunition should be carefully removed, and the gun should be cleaned and oiled according to the manufacturer's instructions. Careful removal from the firearm is not only important to prevent accidents but also to prevent physical damage to the ammunition, making it prone to poor feeding or misfire.

After cleaning the gun, but before reloading it, wipe each round thoroughly with a CLEAN, DRY cloth. DO NOT use the oily rags left from cleaning the gun. NEVER oil ammunition or put it into an oily chamber.

Where rimfire ammunition is used for a law-enforcement purpose, it is recommended that ammunition in the firearm be replaced every six months or sooner. The old ammunition, removed from the firearm, may be used for practice.

3. CONTROL OF AMMUNITION STORED IN ORIGINAL BOXES OR CONTAINERS

The shelf life of rimfire ammunition stored under proper conditions of temperature and humidity, i.e., normal ambient temperatures and humidities in the United States, is well over ten years.

It is suggested that ammunition to be used for security or law-enforcement purposes be replaced every five years or sooner. The old ammunition may be used for practice.

4. CONTROL OF AMMUNITION CARRIED IN BELT LOOPS, CARTRIDGE BOXES, OR CONTAINERS ON BELTS

While rimfire ammunition is not normally carried in belt loops or cartridge boxes, if such a situation does exist it should be noted that ammunition carried in this manner is subjected to the effects of oil from the belts as well as rain or other atmospheric conditions, such as salt air, high humidity, ammonia, etc.

Although pistol and revolver ammunition is usually treated to minimize the contaminating effects of oil solvents or water, prolonged exposure may have an adverse effect on performance.

It is suggested that belt loops and cartridge boxes be unloaded at weekly intervals and the ammunition examined for signs of corrosion or other deterioration. It should be wiped with a

CLEAN, DRY cloth before returning to the belt loops or cartridge box.

For active duty, security, or law-enforcement use, it is recommended that loops or boxes be reloaded at six-month intervals with properly stored ammunition not previously carried on the person. The old ammunition removed from the loops and boxes may be used for practice.

CENTERFIRE PISTOL AND REVOLVER

1. GENERAL

When stored for periods of ten or more years, modern centerfire pistol and revolver ammunition is quite stable under conditions of moderate temperatures and humidity such as are found in the United States.

High temperatures coupled with high humidities over long periods tend to cause some loss in velocity. This result is permanent and cannot be reversed even if the ammunition were held at a constant room temperature for several hours. Long storage at low temperatures will not usually alter anticipated performance *IF* the ammunition is permitted to reach more reasonable temperatures before use. Ammunition fired at temperatures below 0° F. (–18° C.) or above 120° F. (49° C.) may suffer some loss in performance.

Ammunition should never be stored in a corrosive atmosphere, such as in or near chemical plants, stables, the seaside, etc.

2. CONTROL OF AMMUNITION CARRIED IN PISTOLS OR REVOLVERS IN HOLSTERS

Ammunition carried in revolvers or pistols in holsters is subjected to the effects of oil and solvents in the chambers and on the breech face as well as rain or other atmospheric conditions, such as salt air, high humidity, ammonia, etc.

Although pistol and revolver ammunition is usually treated to minimize the contaminating effects of oil, solvents, or water, prolonged exposure may have an adverse effect on performance. Therefore, it is recommended that handgun chambers and magazines be dry to minimize possible contamination of the ammunition by oil or solvents (or both).

Many users, such as law-enforcement officers, are obliged "to store" ammunition in their pistols and revolvers. Here temperature is not such an important factor as are atmospheric conditions, i.e., humidity, corrosive fumes, etc. Situations have been recorded where ammunition has corroded in a firearm because the firearm and ammunition were not properly cleaned over an extended period of time.

It is suggested that law-enforcement and security personnel carefully unload their pistols and revolvers daily, or at least weekly, and examine the ammunition for signs of corrosion or other deterioration, such as green, black, or white discoloration on primer, case, or bullet. Any such evidence recommends removal of the ammunition from service. Careful removal of the rounds from the firearm is not only important to prevent accidents but also to prevent physical damage to the ammunition, making it prone to poor feeding or misfire. The ammunition should be wiped with a CLEAN, DRY cloth before reloading.

Under active-duty conditions, it is recommended that pistols or revolvers be carefully reloaded with fresh, properly stored ammunition at six-month intervals. Ammunition removed from the handgun may be used for practice.

3. CONTROL OF AMMUNITION STORED IN ORIGINAL BOXES OR CONTAINERS

The shelf life of centerfire pistol and revolver ammunition stored under proper conditions of temperature and humidity, i.e., normal ambient temperatures and humidities in the United States, is well over ten years.

It is suggested that ammunition to be used for security or law-enforcement purposes be replaced every five years or sooner. The old ammunition may be used for practice.

4. CONTROL OF AMMUNITION CARRIED IN BELT LOOPS, CARTRIDGE BOXES, OR CONTAINERS ON BELTS

Ammunition carried in belt loops or in cartridge boxes is subjected to the effects of oil from the belts as well as rain or other atmospheric conditions, such as salt air, high humidity, ammonia, etc.

Although pistol and revolver ammunition is usually treated to minimize the contaminating effects of oil, solvents, or water, prolonged exposure may have an adverse effect on performance. This is true for "oil-proof" type ammunition as well.

It is suggested that belt loops and cartridge boxes be unloaded at weekly intervals and the ammunition examined for signs of corrosion or other deterioration. It should be wiped with a CLEAN, DRY cloth before returning to the belt loops or cartridge box.

For active duty, security, or law-enforcement use, it is recommended that loops or boxes be reloaded at six-month intervals with properly stored ammunition not previously carried on the person. The old ammunition removed from the loops and boxes may be used for practice.

CENTERFIRE RIFLE

1. GENERAL
When stored for periods of ten or more years, modern centerfire-rifle ammunition is quite stable under conditions of moderate temperatures and humidity such as are found in the United States.

High temperature coupled with high humidities over long periods tend to cause some loss in velocity. This result is permanent and cannot be reversed even if the ammunition were held at a constant room temperature for several hours. Long storage at low temperatures will not usually alter anticipated performance *IF* the ammunition is permitted to reach more reasonble temperatures before use. Ammunition fired at temperatures below 0° F. (–18° C.) or above 120° F. (49° C.) may suffer some loss in performance.

Ammunition should never be stored in a corrosive atmosphere, such as in or near chemical plants, stables, the seaside, etc.

2. CONTROL OF AMMUNITION CARRIED IN RIFLES
Ammunition carried in rifles is subject to the effects of oil and solvents in the firearms as well as rain or other atmospheric conditions such as salt air, high humidity, ammonia, etc.

Although centerfire rifle ammunition is usually treated to minimize the contaminating effects of oil, solvents, or water, prolonged exposure may have an adverse effect on performance. Therefore, it is recommended that chambers and magazines be dry to minimize possible contamination of the ammunition by oil or solvents (or both).

Ammunition carried in rifles should be removed carefully at weekly intervals and examined for signs of corrosion, discoloration, or deterioration, such as green, black, or white discoloration on primer, case, or bullet. Any such evidence recommends removal of the ammunition from service. Careful removal of the rounds from the firearm is not only important to prevent accidents but also to prevent physical damage to the ammunition, making it prone to poor feeding or misfire. The rounds should be wiped with a CLEAN, DRY cloth before reloading.

Under active-duty conditions, it is recommended that rifles be carefully reloaded with fresh, properly stored ammunition at six-month intervals. Ammunition removed from the rifle may be used for practice.

3. CONTROL OF AMMUNITION STORED IN ORIGINAL BOXES OR CONTAINERS

The shelf life of centerfire rifle ammunition stored under proper conditions of temperature and humidity, i.e., normal ambient temperatures and humidities in the United States, is well over ten years.

It is suggested that ammunition to be used for security or law-enforcement purposes be replaced every five years or sooner. The old ammunition may be used for practice.

4. CONTROL OF AMMUNITION CARRIED IN BELT LOOPS, CARTRIDGE BOXES, OR CONTAINERS ON BELTS

Ammunition carried in belt loops or in cartridge boxes is subjected to the effects of oil from the belts as well as rain or other atmospheric conditions, such as salt air, high humidity, ammonia, etc.

Although rifle ammunition is usually treated to minimize the contaminating effects of oil, solvents, or water, prolonged exposure may have an adverse effect on performance.

It is suggested that belt loops and cartridge boxes be unloaded at weekly intervals and the ammunition examined for signs of corrosion or other deterioration. It should be wiped with a CLEAN, DRY cloth before returning to the belt loops or cartridge box.

For active duty, security, or law-enforcement use, it is recommended that loops or boxes be reloaded at six-month intervals with properly stored ammunition not previously carried on the person. The old ammunition removed from the loops and boxes may be used for practice.

SHOTSHELLS

1. GENERAL

When stored for periods of ten or more years, modern shotshells are quite stable under conditions of moderate temperatures and humidity such as are found in the United States.

High temperatures coupled with high humidities over long periods tend to cause some loss in velocity. This result is permanent and cannot be reversed even if the ammunition were held at a constant room temperature for several hours. Long storage at low temperatures will not usually alter anticipated performance *IF* the ammunition is permitted to reach more reasonable temperatures before use. Ammunition fired at temperatures below 0° F. (–18° C.) or above 120° F. (49° C.) may suffer some loss in performance.

In addition to the adverse effects mentioned above, shotshells with paper tubes may swell under conditions of high temperatures and humidity and may become too large to enter the chamber of the shotgun without force. Shotshells may also become scuffed by excessive loading and unloading in shotguns to the extent that they will not function freely through the action. These conditions can usually be determined by careful, visual examination. Shells exhibiting them may be used for practice or scrapped.

Shotshells should never be stored in a corrosive atmosphere, such as in or near chemical plants, stables, the seaside, etc. Some modern shotshells are made with copperplate steel (instead of brass heads) and should be watched for signs of corrosion.

2. CONTROL OF AMMUNITION CARRIED IN SHOTGUNS

Shotshells carried in shotguns are subject to the effects of oil and solvents in the firearm as well as rain or other atmospheric conditions, such as salt air, high humidity, ammonia, etc.

Shotshells carried in shotguns should be removed carefully at weekly intervals and examined for signs of corrosion or deterioration, such as green, black, or white discoloration on primer or shell head. Any such evidence recommends the removal of the ammunition from service.

Careful removal of the shells from the firearm is not only important to prevent accidents, but also to prevent physical damage to the ammunition, making it prone to poor feeding or misfire. The shells should be wiped thoroughly with a CLEAN, DRY cloth before reloading.

Under active-duty conditions, it is recommended that shotguns be carefully reloaded with fresh, properly stored ammunition at six-month intervals. Shells removed from the shotguns may be used for practice.

3. CONTROL OF AMMUNITION STORED IN ORIGINAL BOXES OR CONTAINERS

The shelf life of shotshells stored under proper conditions of temperature and humidity, i.e., normal ambient temperatures and humidities in the United States, is well over ten years.

It is suggested that ammunition to be used for security or law-enforcement purposes be replaced every five years or sooner. The old ammunition may be used for practice.

4. CONTROL OF AMMUNITION CARRIED IN BELT LOOPS, CARTRIDGE BOXES, OR CONTAINERS ON BELTS

Shotshells carried in belt loops or in cartridge boxes are subject to the effects of oil from the belts as well as rain or other atmospheric conditions, such as salt air, high humidity, ammonia, etc.

Although shotshells are usually treated to minimize the contaminating effects of oil, solvents, or water, prolonged exposure may have an adverse effect on performance.

It is suggested that belt loops and boxes be unloaded at weekly intervals and the shotshells examined for signs of corrosion or deterioration. They should be wiped with a CLEAN, DRY cloth before returning to the belt loop or box.

For active duty, security, or law-enforcement use, it is recommended that loops or boxes be reloaded at six-month intervals with properly stored ammunition not previously carried on the person. The old ammunition removed from the loops and boxes may be used for practice.

CHAPTER SIXTEEN

Historical Exterior Ballistics Tables

Old Federal shotshell boxes.

The information on exterior ballistics tables presented in this chapter were taken from an early Winchester Sales Guide and three early Winchester catalogues.

These tables will be of special interest to the ammunition historian, the cartridge collector, and, of course, to those whose curiosity may be aroused by their presentation.

It is indeed interesting to compare the ballistics of the cartridges of the early 1900's to those that are popular today. Many of the cartridges which have been discontinued for a great number of years took game cleanly and quickly in the hands of the skilled woodsman of yesteryear. You will find listed a great many of the old favorites including the 6mm Lee Navy, .33 Winchester, .38-55 High Velocity, .38-56 Winchester, .38 Express, .40-82 Winchester, .45-70-350, .50-90 Winchester Express, .50-110-300 Winchester High Velocity, and many, many more.

The tables A and B are from the #66 Winchester catalogue dated 1900. Tables C and D are from the #81 Winchester catalogue dated 1918. Tables E and F are from the #83 Winchester catalogue dated 1925. Tables G and H are taken from "Sales Guide of Winchester Metallic Ammunition and Shotshells," which was printed in the early 1940's. The author wishes space had allowed the reprinting of more of this historical data. He also regrets that neither he nor Winchester Press is in the position to supply additional data. For further reference we suggest that you contact local cartridge and/or catalogue collectors.

It is interesting to note that some of the tables are carried out to tenths of a foot pound or to a second.

TABLE A

Showing the Velocity, Penetration, and Trajectory of Winchester Bullets

Name of Rifle Used	Length of Barrel (Inches)	Name of Cartridge	Weight of Bullet (Grains)	Velocity of Bullets (Feet per sec.)	Penetration of Bullets in dry pine boards ⅞ inch thick at 15 feet from muzzle — Plain Lead Boards	Metal Patched Boards	100-Yard Trajectory, Height at 50 Yards (Inches)	200-Yard Trajectory, Height at 100 Yards (Inches)	300-Yard Trajectory, Height at 150 Yards (Inches)
Model 1890	24	.22 Winchester R.F.	45	1137	4		4.0	12.6	33.7
Single Shot	26	.22 Winchester S.S.	45	1481	5		2.7		
Lee Stgt. Pull	28	6mm U.S. Navy	112	2500		60	.8	3.6	9.4
Model 1892	24	.25-20 W.C.F.	86	1300	9		3.3	13.8	34.7
Single Shot	28	.25-20	86	1304	9		3.4	13.6	34.7
Model 1894	26	.25-35 W.C.F.	117	1925		36	2.4	5.1	13.9
Model 1894	26	.30 W.C.F.	160	1885		35	2.1	5.2	13.6
Savage Rifle	26	.303 Savage	180	1840		33	1.2	6.3	16.4
Model 1895	28	.30 U.S. Army	220	1960		58	1.5	5.1	14.1
Model 1892	24	.32 Winchester	115	1177	6½		3.5	15.4	37.2
Single Shot	30	.32-40	165	1385	8½	18	2.7	11.3	28.3
Model 1892	24	.38 Winchester	180	1268	7½		3.2	14.4	35.7
Single Shot	30	.38-55	255	1285	9½		3.0	12.9	32.0
Express S.S.	30	.38-90 Win. Exp.	217	1546	9	17	2.2	8.6	22.8
Model 1886	26	.38-56 Winchester	255	1359	11	14½	2.8	12.2	30.1
Model 1886	26	.38-70 Winchester	255	1449	10	19	2.6	10.6	27.2
Model 1895	26	.38-72 Winchester	275	1443	16	25	2.2	10.6	27.7
Single Shot	26	.40-70 Sharp's Stgt.	330	1229	11½		3.3	13.4	32.9
Marlin Rifle	28	.40-60 Marlin	260	1419	8½		3.0	11.8	29.4
Single Shot	28	.40-60 Winchester	210	1475	9½		2.6	11.7	30.1

TABLE B

Showing the Velocity, Penetration, and Trajectory of Winchester Bullets

| Name of Rifle Used | Length of Barrel | Name of Cartridge | Weight of Bullet | Velocity of Bullets | Penetration of Bullets in dry pine boards ⅞ inch thick at 15 feet from muzzle | | Trajectories of Bullets | | |
					Plain Lead Boards	Metal Patched Boards	100-Yard Trajectory Height at 50 Yards	200-Yard Trajectory Height at 100 Yards	300-Yard Trajectory Height at 150 Yards
	Inches		Grains	Feet per sec.	Boards	Boards	Inches	Inches	Inches
Model 1895	26	.40-72 Win. B.P.	330	1359	13	23	2.6	12.2	30.5
Model 1895	26	.40-72 Winchester	300	1386		22	2.4	11.6	28.5
Single Shot	30	.40-90 Sharp's Stgt.	370	1357	16	22	2.7	10.8	26.9
Model 1886	26	.40-65 Winchester	260	1325	9	14½	2.9	12.0	30.7
Model 1886	26	.40-70 Winchester	330	1349	13	19½	2.8	11.8	29.4
Model 1886	26	.40-82 Winchester	260	1445	12	17½	2.6	11.9	30.3
Express S.S.	30	.40-110 Win. Exp.	260	1555	12¼		2.1	9.0	23.6
Model 1892	24	.44 Winchester	200	1245	9		3.4	15.3	37.4
Single Shot	30	.45-75 Winchester	350	1343	14½		3.0	12.4	30.6
Single Shot	30	.45-60 Winchester	300	1271	11½		3.2	13.7	33.1
Model 1886	26	.45-70-500	500	1179	18		3.7	14.4	34.4
Model 1886	26	.45-70-405	405	1271	14	16½	3.3	13.1	32.4
Model 1886	26	.45-70-405 Smokeless	405	1286		20	4.1	12.3	29.0
Model 1886	26	.45-70-350 Win.	350	1307	13		2.8	13.1	32.4
Model 1886	26	.45-70-330 Gould	330	1338	10		2.8	12.7	31.8
Model 1886	26	.45-90 Winchester	300	1480	13	19	2.4	10.3	27.3
Express S.S.	30	.45-125 Win. Exp.	300	1633	9½		2.2	9.0	25.1
Model 1886	26	.50-110 Win. Exp.	300	1536	11		2.5	11.9	33.5
Model 1886	26	.50-100-450 Win.	450	1383	16		2.9	11.9	30.7
Express S.S.	30	.50-95 Win. Exp.	300	1493	10		2.6	12.6	33.5

TABLE C

Velocities, Energies, Penetrations and Trajectories Developed by Winchester Cartridges and Recoils of Winchester Rifles

Penetration is not the measure of striking energy. As an illustration, take the figures in our table for the .30 Winchester Center Fire Cartridge. With the soft point bullet the penetration is but 11 boards, whereas that cartridge with the full patch bullet will penetrate 42 boards. The energies of both are the same. All other things being equal, the bullet which resists deformation will give the maximum penetration. The soft point bullet, which generally stops inside the skin of the animal, delivers its whole energy; while the full patch bullet, which passes through the animal, may make a less severe wound. Penetration, therefore, is not a good test of killing power. If the target is harder or softer than that described in our table, the results obtained will not be the same; nor will the comparative results show corresponding differences. The rifles used in determining the following ballistics were equipped with barrels of standard lengths.

Name of Cartridge	Weight of Bullets	Velocities of Bullets Ft. per Sec.		Energies of Bullets in Ft. Lbs.		Penetrations at 15 Ft. in ⅞ Inch Soft Pine Boards			Trajectories			Free Recoil in Ft. Lbs.	
	Grains	Muzzle	100 Yds.	Muzzle	100 Yds.	Lead	S.P.	F.M.C.	100 Yds. Ht. at 50 Yds.	200 Yds. Ht. at 100 Yds.	300 Yds. Ht. at 150 Yds.	Smokeless	Black Powder
6 m/m	112	2562.0	2231.5	1632.8	1239.1		12	60	.8	3.5	9.1	7.1	
.22 Win. C.F.	45	1541.2	1126.0	237.4	126.7	8			2.6	13.7	38.3	.4	.5
.25-20 S.S.	86	1411.9	1132.6	380.8	245.0	9	8	11	2.7	13.5	35.8	.5	.7
.25-20 Winchester	86	1376.3	1108.6	361.8	234.7	9	8	11	2.9	14.1	41.0	.8	.8
.25-20 W.H.V.	86	1728.1	1407.9	570.4	378.6		10	20	1.9	8.9	24.3	1.4	
.25-35 Winchester	117	1973.0	1698.1	1011.6	749.6		11	36	1.3	6.1	16.0	3.4	
.30 Winchester	170	2003.4	1753.0	1515.5	1160.2		11	42	1.2	5.7	14.8	7.2	
.30 Army Ptd.	150	2557.9	2331.7	2179.8	1811.3				.7	3.3	8.2		
.30 Army Ptd.	180	2345.5	2167.4	2199.3	1878.0				.9	3.7	9.1		
.30 Army	220	1993.5	1798.4	1941.8	1580.3		13	58	1.2	5.4	13.5	11.6	
.30 Govt.'03	220	2198.9	1989.7	2362.5	1934.3		18	68	1.0	4.4	11.0	15.0	
.30-06 Gov't.	150	2700.0	2465.1	2428.6	2024.5		14	75	.6	2.9	7.9	11.4	
.30-06 Gov't.	180	2499.4	2313.1	2497.5	2139.1		18		.8	3.3	7.9	12.6	
.30-06 Gov't.	220	2198.9	1989.7	2362.5	1934.3		13	56	1.0	4.4	11.0	15.0	
.303 British	215	1999.1	1775.7	1908.3	1505.7		10	17	1.2	5.5	14.1	11.0	
.32 Win. S.L.	165	1392.0	1167.0	710.1	499.1		6.5	10	2.7	12.5	33.3	1.9	
.32 Winchester	115	1222.2	1010.9	381.5	261.0	6.5			3.6	16.9	43.6	1.1	1.2
.32 Win. Special	170	2104.4	1792.7	1672.0	1213.5		12	45	1.2	5.6	14.6	7.7	
.32 W.H.V.	115	1636.0	1304.3	683.6	434.5		7	17	2.0	10.4	28.0	2.7	

TABLE C (continued)

Name of Cartridge	Weight of Bullets (Grains)	Velocities of Bullets Ft. per Sec.		Energies of Bullets in Ft. Lbs.		Penetrations at 15 Ft. in % Inch Soft Pine Boards			Trajectories			Free Recoil in Ft. Lbs.	
		Muzzle	100 Yds.	Muzzle	100 Yds.	Lead	S.P.	F.M.C.	100 Yds. Ht. at 50 Yds.	200 Yds. Ht. at 100 Yds.	300 Yds. Ht. at 150 Yds.	Smoke-less	Black Powder
.32-40	165	1427.7	1194.5	747.0	522.9	8.5	8.5	18	2.5	12.2	31.9	3.1	4.1
.32-40 W.H.V.	165	1748.5	1476.9	1120.4	799.4		10	30	1.7	8.1	21.5	5.5	
.33 Winchester	200	2050.3	1761.7	1867.3	1378.7		13	39	1.2	5.6	15.0	11.4	
.35 Win. S.L.	180	1396.0	1151.0	779.1	529.6		9	17	2.7	13.1	34.4	2.8	
.35 Winchester	250	2192.7	1945.3	2669.6	2101.1		15	56	1.0	4.6	11.8	19.8	
.351 Win. S.L.	180	1856.4	1541.7	1377.8	950.1		13	26	1.5	7.4	20.2	5.6	

TABLE D

Velocities, Energies, Penetrations and Trajectories Developed by Winchester Cartridges and Recoils of Winchester Rifles

Name of Cartridge	Weight of Bullets (Grains)	Velocities of Bullets Ft. per Sec.		Energies of Bullets in Ft. Lbs.		Penetrations at 15 Ft. in % Inch Soft Pine Boards			Trajectories			Free Recoil in Ft. Lbs.	
		Muzzle	100 Yds.	Muzzle	100 Yds.	Lead	S.P.	F.M.C.	100 Yds. Ht. at 50 Yds.	200 Yds. Ht. at 100 Yds.	300 Yds. Ht. at 150 Yds.	Smoke-less	Black Powder
.38 Winchester	180	1324.0	1053.3	700.8	443.5	7.5	10	12	3.2	15.5	41.7	3.2	4.7
.38 W.H.V.	180	1770.0	1389.6	1252.5	771.9		10	20	1.8	9.3	25.6	6.7	
.38-55	255	1321.0	1131.6	988.3	725.2	9.5	13.5	14	2.9	13.6	34.4	6.0	8.4
.38-55 W.H.V.	255	1590.1	1364.2	1432.0	1054.0		10	23	2.0	9.4	24.4	9.4	
.38-56 Winchester	255	1397.0	1189.2	1105.3	800.9	11	12	17	2.6	12.3	31.0	5.8	8.1
.38-70 Winchester	255	1489.5	1262.9	1256.5	903.3	10	12	19	2.1	11.8	28.7	7.2	10.2
.38-72 Winchester	275	1476.6	1286.1	1331.7	1010.2	16	15	25	2.3	10.6	27.1	8.7	9.4
.38 Express	217	1595.8	1813.7	1227.4	831.8	9			2.1	10.2	27.5		9.6
.40-60 Winchester	210	1532.7	1220.3	1095.7	694.6	9.5			2.3	11.8	32.0		6.9
.40-65 Winchester	260	1367.2	1145.1	1079.4	757.7	9	11	14.5	2.6	13.2	33.4	6.8	8.7

.40-70 Winchester	330	1382.8	1196.7	1401.5	1049.6	13	11	19.5	2.7	12.2	33.9	9.2	13.0
.40-72 Winchester	330	1373.0	1190.6	1381.6	1063.2	13	14	22	2.8	12.1	33.9	10.0	14.6
.40-72 Winchester	300	1423.8	1214.6	1350.7	983.0	12	11	17.5	2.5	11.8	30.6	8.8	12.2
.40-82 Winchester	260	1492.1	1236.9	1285.6	883.5		14	34	2.4	11.3	29.9	11.5	
.401 Win. S.L.	200	2132.7	1749.2	2020.3	1359.1		12	27	1.2	5.8	16.1	12.2	
.401 Win. S.L.	250	1869.9	1562.2	1941.5	1355.1		13	48	1.5	7.2	19.6	28.2	
.405 Winchester	300	2197.5	1923.1	3217.6	2464.2	9	10	13	1.0	4.7	12.3	3.9	5.4
.44 Winchester	200	1300.6	1034.6	751.4	475.5		10	19	3.3	15.9	42.4	6.0	
.44 W.H.V.	200	1563.9	1226.1	1086.3	667.8	11.5			2.3	11.6	31.2		9.3
.45-60 Winchester	300	1314.6	1091.8	1151.5	794.2				3.0	14.5	37.4		
.45-70 W.H.V.	300	1882.9	1559.0	2362.4	1619.5	13	13	25	1.5	7.2	19.8	16.2	14.6
.45-40-350	350	1343.8	1139.1	1403.8	1008.6	13	11	17	2.9	15.0	34.3	10.3	16.2
.45-70-405 Govt.	405	1317.6	1143.3	1561.7	1175.8	18	12	18	2.9	13.3	33.6	12.3	18.4
.45-70-500 Govt.	500	1201.1	1081.6	1602.1	1317.2	14.5	15	20	3.5	14.8	36.1	15.2	13.6
.45-75 Winchester	350	1382.7	1168.2	1485.1	1060.8	13			2.7	13.0	32.9		16.5
.45-90 Winchester	300	1531.7	1247.8	1563.3	1037.5		15	19	2.3	11.2	30.2	11.4	
.45-90 W.H.V.	300	1985.7	1643.2	2627.2	1798.7	10	14	26	1.3	6.5	18.0	19.0	17.5
.50-95 Win. Exp.	300	1556.8	1214.2	1614.8	982.3	16			2.3	11.9	32.7		25.2
.50-100-450 Win.	450	1422.1	1206.6	2021.2	1455.2	11	14	20	2.5	12.0	32.1	21.5	19.8
.50-110 Win. Exp.	300	1605.8	1250.2	1718.2	1041.4		12	20	2.2	11.0	31.2	11.3	
.50-110-300 W.H.V.	300	2230.1	1779.4	3313.7	2109.8		14	26	1.1	5.6	16.2	25.6	

TABLE E

Velocities, Energies, Penetrations and Trajectories Developed by Winchester Cartridges and Recoils of Winchester Rifles

Penetration is not the measure of striking energy. As an illustration, take the figures in our table for the .30 Winchester Center Fire Cartridge. With the soft point bullet the penetration is but 11 boards, whereas that cartridge with the full patch bullet will penetrate 50 boards. The energies of both are the same. All other things being equal, the bullet which resists deformation will give the maximum penetration. The soft point bullet, which generally stops inside the skin of the animal, delivers its whole energy; while the full patch bullet, which passes through the animal, may make a less severe wound. Penetration, therefore, is not a good test of killing power. If the target is harder or softer than that described in our table, the results obtained will not be the same; nor will the comparative results show corresponding differences. The rifles used in determining the following ballistics were equipped with barrels of standard lengths.

Name of Cartridge	Weight of Bullets Grains	Velocities of Bullets Ft. per Sec.		Energies of Bullets in Ft. Lbs.		Penetrations at 15 Ft. in ⅞ Inch Soft Pine Boards			Trajectories			Free Recoil in Ft. Lbs.	
		Muzzle	100 Yds.	Muzzle	100 Yds.	Lead	S.P.	F.M.C.	100 Yds. Ht. at 50 Yds.	200 Yds. Ht. at 100 Yds.	300 Yds. Ht. at 150 Yds.	Smokeless	Black Powder
6mm	112	2560	2230	1635	1240		12	60	.7	3.5	9.0	7.0	
.22 Win. C.F.	45	1540	1125	240	125	8			2.5	13.5		.4	.5
.25-20 S.S.	86	1410	1135	380	245	9	8	11	2.5	13.5		.5	.7
.25-20 Winchester	86	1375	1110	360	235	9	8	11	3.0	14.0		.8	.9
.25-20 W.H.V.	86	1730	1410	570	380		10	20	2.0	9.0			
.25-35 Winchester	117	2175	1880	1230	920		11	44	1.0	5.0	13.0	1.5	
.270 Winchester	130	3160	2970	2880	2550		17		0.5	2.0	4.5	4.5	
.30 Winchester	170	2200	1930	1825	1410		11	50	1.0	4.5	12.0		
.30 Army, Ptd.	150	2560	2330	2180	1810				.7	3.5	8.0	9.0	
.30 Army, Ptd.	180	2345	2165	2200	1880				.8	3.5	9.0		
.30 Army	220	1995	1800	1940	1580		13	58	1.0	5.5	13.0	11.5	
.30 Gov't, '03	220	2200	1990	2365	1935		18	68	1.0	4.5	11.0	15.0	
.30-06 Gov't	150	2700	2465	2430	2025		14	75	.6	3.0	7.0	11.5	
.30-06 Gov't	180	2500	2315	2500	2140				.7	3.5	8.0	12.5	
.30-06 Gov't	180	2700	2505	2915	2505		17		.5	2.5	7.0	16.5	
.30-06 Gov't	220	2400	2185	2810	2340		17		1.0	4.0	9.0	16.0	
.303 British	215	2000	1775	1910	1505		13	56	1.0	5.5	14.0	11.0	
.32 Win. S.L.	165	1390	1165	710	500		10	17	2.5	12.5	33.0	2.0	

TABLE F
Velocities, Energies, Penetrations and Trajectories Developed by Winchester Cartridges and Recoils of Winchester Rifles

Name of Cartridge	Weight of Bullets	Velocities of Bullets Ft. per Sec.		Energies of Bullets in Ft. Lbs.		Penetrations at 15 Ft. in ⅞ Inch Soft Pine Boards			Trajectories			Free Recoil in Ft. Lbs.	
	Grains	Muzzle	100 Yds.	Muzzle	100 Yds.	Lead	S.P.	F.M.C.	100 Yds. Ht. at 50 Yds.	200 Yds. Ht. at 100 Yds.	300 Yds. Ht. at 150 Yds.	Smoke-less	Black Powder
.32 Winchester	115	1225	1010	380	260	7	7	10	3.5	17.0		1.1	1.2
.32 Win. Spl.	170	2250	1925	1910	1395		12	52	1.0	4.5	12.0	9.5	
.32 W.H.V.	115	1635	1305	685	435		7	17	2.0	10.5	28.0	2.5	
.32-40	165	1430	1195	745	525	9	9	18	2.5	12.0	32.0	3.0	4.1
.32-40 W.H.V.	165	1750	1475	1120	800		10	30	1.5	8.0	21.0	5.5	
.33 Winchester	200	2200	1895	2150	1580		13	45	1.0	5.0	13.0	13.5	
.35 Win. S.L.	180	1395	1150	780	530		9	17	2.5	13.0	34.0	3.0	
.35 Winchester	250	2195	1945	2670	2100		15	56	1.0	4.5	12.0	20.0	
.351 Win. S.L.	180	1855	1540	1380	950		13	26	1.5	7.5	20.0	5.5	
.38 Winchester	180	1325	1055	700	445	8	10	12	3.0	15.5	41.0	3.0	4.5
.38 W.H.V.	180	1770	1390	1255	770		10	20	2.0	9.5	25.0	6.5	
.38-55	255	1320	1130	990	725	10	14	14	3.0	13.5	34.0	6.0	8.5
.38-55 W.H.V.	255	1590	1865	1430	1055		10	23	2.0	9.5	24.0	9.5	
.38-56 Winchester	255	1395	1190	1105	800	11	12	17	2.5	12.5	32.0	5.5	8.0
.38-72 Winchester	275	1475	1285	1330	1010	16	15	25	2.5	10.5	27.0	8.5	9.5
.40-60 Winchester	210	1533	1220	1095	695	10			2.5	11.5	32.0		7.0
.40-65 Winchester	260	1370	1145	1080	760	9	11	15	2.5	13.0	33.0	7.0	8.5
.40-72 Winchester	300	1425	1215	1350	985		14	22	2.5	12.0	30.0	10.0	
.40-82 Winchester	260	1490	1235	1285	885	12	11	18	2.5	11.0	30.0	9.0	12.0
.401 Win. S.L.	200	2135	1750	2020	1360		14	34	1.0	5.5	16.0	11.5	
.401 Win. S.L.	250	1870	1560	1940	1355		12	27	1.5	7.0	19.0	12.0	
.405 Winchester	300	2200	1925	3220	2465		13	48	1.0	4.5	12.0	28.0	
.44 Winchester	200	1300	1035	750	475	9	10	13	3.0	16.0	42.0	4.0	5.5
.44 W.H.V.	200	1565	1225	1085	670		10	19	2.5	11.5	31.0	6.0	
.45-60 Winchester	300	1315	1090	1150	795	12			3.0	14.5	37.0		9.5
.45-70 W.H.V.	300	1885	1560	2365	1620		13	25	1.5	7.0	20.0	16.0	

TABLE F (continued)

Velocities, Energies, Penetrations and Trajectories Developed by Winchester Cartridges and Recoils of Winchester Rifles

Name of Cartridge	Weight of Bullets	Velocities of Bullets Ft. per Sec.		Energies of Bullets in Ft. Lbs.		Penetrations at 15 Ft. in ⅞ Inch Soft Pine Boards			Trajectories			Free Recoil in Ft. Lbs.	
	Grains	Muzzle	100 Yds.	Muzzle	100 Yds.	Lead	S.P.	F.M.C.	100 Yds. Ht. at 50 Yds.	200 Yds. Ht. at 100 Yds.	300 Yds. Ht. at 150 Yds.	Smoke-less	Black Powder
.45-70-405 Govt.	405	1320	1145	1560	1175	13	12	18	3.0	13.0	33.0	12.0	16.0
.45-75 Winchester	350	1385	1170	1495	1060	15			3.0	13.0	33.0		13.5
.45-90 Winchester	300	1530	1250	1565	1040	13	15	19	2.5	11.0	30.0	11.5	16.5
.45-90 W.H.V.	300	1985	1645	2630	1800		14	26	1.5	6.5	18.0	19.0	
.50-110 Win. Exp.	300	1605	1250	1720	1040	11	13	20	2.0	11.0	31.0	11.5	19.5

TABLE G

Rifle Sighting Tables for Winchester Cartridges

To help the shooter to quickly adapt his holding to various ranges, Winchester has developed the following tables showing the approximate actual positions of the bullets at the ranges given. These are based on the rimfire rifle being zeroed at 50 yards and the centerfire rifle at 100 yards—in many cases 200 yards also. These figures show the distance in inches above or below the line of sight. In the centerfire tables positions above the line of sight are indicated by a plus sign, those below by a minus sign. It must be understood, of course, that due to wind conditions and other factors there are bound to be variations in the flight of bullets, which cannot be avoided. Therefore, these figures in all cases must be read as indicating *approximate* positions. The rifles used in determining these ballistics had standard length barrels. This new presentation has been developed because it makes it simpler and quicker for the shooter to adapt his holding to any ranges.

Cartridge	Bullet Type	Wt.-Grs.	100 Yds.	Drop of Bullet Zero at 50 Yards 150 Yds.	200 Yds.
Leader .22 Short	Lead	29	9.5"	—	—
Super Speed .22 Short	K.K.	29	7.0"	31.0"	67.0"
Super Speed .22 Short	K.K., H.P.	27	7.0"	31.5"	68.0"
Leader .22 Long	Lead	29	8.0	—	—
Super Speed .22 Long	K.K.	29	5.5	24.5	54.0
Super Speed .22 Long	K.K., H.P.	27	6.0	25.0	58.0
Leader .22 Long Rifle	Lead	40	6.5	—	—
EZXS, .22 Long Rifle	Lead	40	7.5	—	—
All-X Match .22 Long Rifle	Lead	40	6.5	—	—
Super Speed .22 Long Rifle	K.K.	40	5.0	24.0	50.5
Super Speed .22 Long Rifle	K.K., H.P.	37	5.5	22.5	50.0
.22 Extra Long	Lead	40	9.0	—	—
.22 W.R.F.	Lead	45	6.5	—	—
.22 W.R.F.	K.K.	45	6.5	—	—
Super Speed .22 W.R.F.	K.K.	45	5.0	22.5	46.0
Super Speed .22 W.R.F.	K.K., H.P.	40	5.0	22.5	47.5
.22 Automatic	Lead	45	8.0	—	—
.22 Automatic	K.K.	45	8.0	—	—
.22 Automatic	K.K., H.P.	45	8.0	—	—
.25 Stevens	Lead	65	6.0	—	—
.25 Short Stevens	Lead	65	9.5	—	—
.32 Short	Lead	80	9.0	—	—
.32 Long	Lead	90	8.5	—	—
.41 Swiss	Lead	310	9.0	—	—

K.K. = Kopperklad   H.P. = Hollow Point

TABLE H
Range Table for Winchester Center Fire Cartridges

Path of Bullet Above or Below Line of Sight in Inches

Cartridge	Bullet Wt. Grs.	Type	50 Yards	100 Yards	200 Yards	300 Yards	400 Yards	500 Yards
Super Speed .218 Win. Bee	46	H.P.	+0.7	0	−6.0	−26.5		
Super Speed .219 Win. Zipper	46	H.P.	+0.4	+3.5	−4.0	−19.0		
Super Speed .219 Win. Zipper	56	H.P.	+0.6	+2.5	−4.5	−18.0		
Super Speed .22 Win. Hornet	45	S.P.	+0.8	+2.5	−7.5	−32.0		
Super Speed .22 Win. Hornet	46	H.P.	+0.8	+4.0	−6.5	−31.5		
Super Speed .22 High Power	70	P.S.P.	+0.6	+3.5	−4.5	−17.0		
Super Speed .220 Win. Swift	48	P.S.P.	+0.3	+1.5	−2.5	−9.0	−24.0	−50.5
Super Speed .220 Win. Swift	55	P.S.P.	+0.35	+1.5	−2.5	−11.0	−27.0	−56.0
.25-20 Winchester	86	Lead	+2.6	0	−18.5	−76.0		
.25-20 Winchester	86	F.P.	+2.6	0	−18.5	−76.0		
.25-20 Winchester	86	S.P.	+2.6	0	−18.5	−76.0		
Super Speed .25-20 W.H.V.	86	S.P.	+1.9	0	−15.5	−62.0		
Super Speed .25-20 Win.	60	H.P.	+1.2	0	−11.0	−48.0		
.25-20 Single Shot	86	S.P.	+2.7	0	−22.0	−80.5		
.25 Remington Auto	117	S.P.	+0.9	0	−6.5	−26.0		
6.5 m/m Mannlicher	145	F.P.	+0.9	0	−6.5	−24.5	−58.0	−116.5
Super Speed .25-35 Win.	117	F.P.	+1.0	0	−7.5	−29.5		
Super Speed .25-35 Win.	117	S.P.	+1.0	0	−7.5	−29.5		
Super Speed .257 Win. Roberts	87	P.S.P.	+0.49	+2.0	−3.5	−13.0	−31.0	−60.0
Super Speed .257 Win. Roberts	100	P.E.	+0.59	+2.5	−3.5	−15.0	−36.0	−70.0
Super Speed .250-3000 Savage	87	F.P.	+0.5	+2.5	−3.5	−14.5	−34.5	−69.0

Cartridge	Bullet	Wt.						
Super Speed .250-3000 Savage	S.P.	87	+0.5	0	−3.5	−14.5	−24.5	−69.0
Super Speed .250-3000 Savage	P.E.	100	+0.6	+2.5	−4.5	−9.0	−26.5	−58.0
				+2.5	0	−16.0	−38.5	−73.5
Super Speed .270 Win.	P.E.	130	+0.5	+2.5	−3.5	−10.0	−29.5	−62.0
Super Speed. 270 Win.	S.P.	150	+0.6	+2.0	0	−12.5	−30.0	−54.5
					−7.5	−23.0	−46.0	
Super Speed .270 Win.	P.E.	100	+0.4	+2.5	−4.5	−18.0	−44.0	−86.5
				+1.5	−3.0	−11.0	−33.5	−72.5
Super Speed 7 m/m Mauser	S.P.	175	+0.8	0	−5.5	−22.0	−52.0	−100.0
Super Speed 7 m/m Mauser	P.E.	150	+0.6	0	−4.5	−15.5	−34.5	−67.0
				+2.5		−9.5	−26.0	−56.0
7.62 m/m Russian	H.C.P.	145	+0.6	0	−3.5	−14.0	−40.5	−78.0
Super Speed .30 Win.	F.P.	170	+1.0	0	−7.5	−28.0		
Super Speed .30 Win.	S.P.	170	+1.0	0	−7.5	−28.0		
Super Speed .30 Win.	H.P.	110	+0.7	0	−6.0	−24.5		
Super Speed .30 Win.	H.P.	150	+0.9	0	−6.5	−26.0		
.30 Remington Auto	S.P.	170	+1.1	0	−7.5	−29.5		
Super Speed .300 Savage	P.S.P.	150	+0.7	+3.0	−4.5	−17.5	−41.5	−79.5
					−11.0	−32.0	−67.0	
Super Speed .300 Savage	S.P.	180	+0.9	0	−7.5	−26.5	−58.0	−114.5
.303 Savage	S.P.	190	+1.5	0	−7.5	−35.0	−69.5	−130.0
Super Speed .303 British	S.P.	215	+1.0	0	−7.0	−28.5	−65.5	−127.5
.30 Army (.30-40 Krag)	F.P.	220	+1.0	0	−7.0	−27.0	−65.5	−127.5
.30 Army (.30-40 Krag)	S.P.	220	+1.0	0	−7.0	−27.0	−41.5	−79.5
Super Speed .30 Army Ptd. (.30-40 Krag)	P.E.	150	+0.7	0	−4.0	−17.5	−32.0	−67.0
				+3.0	0	−11.0	−46.0	−87.0
Super Speed .30 Army Ptd. (.30-40 Krag)	P.E.	180	+0.8	+3.5	−4.5	−19.5	−35.0	−73.0
					−11.5	−55.0	−108.0	
Super Speed .30 Army (.30-40 Krag)	S.P.	180	+0.8	0	−5.5	−22.5	−42.5	−92.5
Super Speed .30-06 Gov't.	P.E.	150	+0.5	+3.5	0	−14.0	−33.0	−63.0
				+2.5	−3.5	−14.0	−25.0	−53.0
.30-06 Gov't.	F.P.	180	+0.6	0	−4.5	−16.0	−43.5	−78.5
				+3.0		−9.0	−32.5	−65.0
Super Speed .30-06 Gov't.	P.E.	180	+0.6	+3.0	−4.5	−16.5	−38.0	−71.5
						−9.0	−28.0	−59.0
Super Speed .30-06 Gov't.	S.P.	180	+0.7	+3.0	−4.5	−16.5	−44.5	−88.5
						−9.5	−35.0	−76.0
Super Speed .30-06 Gov't.	S.P.	220	+0.8	0	−5.5	−22.0	−52.0	−102.5

199

Range Table for Winchester Center Fire Cartridges (continued)

Cartridge	Bullet Type	Wt. Grs.	50 Yards	100 Yards	200 Yards	300 Yards	400 Yards	500 Yards
				Path of Bullet Above or Below Line of Sight in Inches				
.30-06 Gov't. Boattail Precision	F.P.	172	+0.6	+3.5	0	-13.5	-36.5	-87.0
				0	-4.5	-16.0	-38.0	-73.0
.30-06 Gov't. Wimbledon Cup Boattail	F.P.	180	+0.6	+3.0	0	-9.5	-28.5	-61.0
					-4.5	-16.5	-38.0	-71.5
.300 H&H Magnum Match	F.P.	180	+0.5	+3.0	0	-9.5	-28.0	-59.0
					-3.5	-12.5	-29.0	-55.0
Super Speed .300 H&H Magnum Boattail	H.P.	180	+0.55	+2.0	0	-7.5	-22.0	-45.5
					-4.5	-15.0	-38.0	-76.0
Super Speed .300 H&H Magnum Boattail	S.P.	220	+0.7	+2.5	-4.5	-9.5	-29.5	-65.0
				+3.0	-4.5	-18.0	-44.5	-86.0
				0	0	-11.0	-34.5	-73.0
8mm (7.9mm)	S.P.	236	+1.1	0	-7.5	-29.0	-70.5	-136.5
Super Speed 8mm Mauser Boattail	S.P.	170	+0.8		-5.5	-23.0	-58.0	-116.0
.32 Winchester	Lead	100	+3.1		-24.5	-98.0		
.32 Winchester	F.P.	115	+3.1		-23.5	-90.5		
.32 Winchester	S.P.	115	+3.1		-23.5	-90.5		
*Super Speed .32 W.H.V.	S.P.	115	+2.2		-17.0	-67.5		
*Super Speed .32 Win.	H.P.	80	+1.3		-13.5	-59.0		
Super Speed .32 Win. Spl.	S.P.	170	+1.0		-7.5	-28.5		
Super Speed .32 Win. Spl.	H.P.	110	+0.8		-5.5	-28.5		
.32 Win. Self-Loading	S.P.	165	+2.6		-21.0	-74.0		
.32 Remington Auto	S.P.	165	+1.0		-7.5	-31.0		
.32-40	S.P.	165	+2.6		-19.5	-66.0		
.33 Winchester	S.P.	200	+1.1		-8.0	-32.5		
Super Speed .348 Win.	S.P.	150	+0.6	+3.0	-5.5	-21.5	-70.0	-143.0
Super Speed .348 Win.	S.P.	200	+0.8	+4.0	-6.5	-14.5	-58.0	-128.0
					0	-24.0	-73.0	-147.5
						-16.0	-58.5	-129.5
.35 Winchester	S.P.	250	+1.1		-7.5	-29.0		
.35 Win. Self-Loading	S.P.	180	+2.5		-22.0	-74.0		
.35 Remington Auto	S.P.	200	+1.1		-8.5	-32.0		
.351 Win. Self-Loading	F.P.	180	+1.5		-12.5	-45.5		
.351 Win. Self-Loading	S.P.	180	+1.5		-12.5	-45.5		
Super Speed .375 H&H Magnum	S.P.	270	+0.7	0	-4.5	-17.5	-41.5	-82.5
Super Speed .375 H&H Magnum	S.P.	300	+0.7	+3.0	0	-10.5	-32.0	-70.5
Super Speed .300 H&H Magnum	P.E.	150	+0.5	0	-5.5	-21.0	-49.0	-94.5
					-3.0	-13.0	-29.0	-57.0

Super Speed .300 H&H Magnum	P.E.	180	+0.5	+2.0	0	-8.0	-22.0	-48.0
				0	-4.5	-13.5	-32.0	-59.5
Super Speed .300 H&H Magnum	H.P.,B.T.	180	+0.55	+2.5	-4.5	-8.0	-24.0	-49.0
				0	0	-15.0	-38.0	-76.0
Super Speed .300 H&H Magnum	S.P.,B.T.	220	+0.7	+2.5	-4.5	-9.5	-29.5	-65.0
						-18.0	-44.5	-86.0
				+3.0	0	-11.0	-34.5	-73.0
.38 Winchester	S.P.	180	+3.3	0	-26.0	-90.5		
* .38 W.H.V.	S.P.	180	+1.7	0	-15.5	-61.0		
.38-55	S.P.	255	+3.0	0	-22.0	-76.0		
.38-56 Winchester	S.P.	255	+2.7	0	-17.5	-67.5		
.38-72 Winchester	S.P.	275	+2.4	0	-15.5	-63.0		
.40-65 Winchester	S.P.	260	+2.9	0	-19.5	-72.0		
.40-82 Winchester	S.P.	260	+2.3	0	-18.0	-64.0		
.401 Win. Self-Loading	S.P.	200	+1.2	0	-9.5	-41.0		
.405 Winchester	S.P.	300	+1.0	0	-7.0	-31.0		
.44 Winchester	S.P.	200	+3.3	0	-31.0	-91.0		
* .44 W.H.V.	S.P.	200	+2.2	0	-19.5	-73.5		
.45-70-405 Gov't.	S.P.	405	+2.8	0	-21.0	-76.0		
.45-90 Winchester	S.P.	300	+2.2	0	-18.5	-63.5		

S.P. = Soft Point   H.P. = Hollow Point   F.P. = Full Patch   P.E. = Pointed Expanding   H.C.P. = Hollow Copper Point

*Not adapted to pistols or revolvers or to Winchester Model 73 rifle.

K.K. = Lead, Kopperklad   H.P. = Hollow Point

CHAPTER SEVENTEEN

Ammunition and Ballistics Glossary

The following glossary lists some of the terms and their definitions as normally used in the ammunition industry. We have endeavored to include those terms most commonly used and those which are often used incorrectly. Dictionary definitions are not included as they are not unique to the ammunition industry. Special thanks are given to SAAMI for supplying the common ground for this glossary.

Accuracy—In firearms using single projectiles, the measure of the dispersion of the group fired. The optimum would be one hole no larger in diameter than a single projectile.

Adapter—A device which permits use of smaller caliber ammunition in a firearm designed for a larger caliber.

Aiming Point—A point on the target upon which the sights are aligned.

Air Resistance—The resistance of air to the passage of a projectile in flight.

Air Space—The volume in a loaded cartridge or shotshell not occupied by the propellant, the bullet, wadding, or shot. Sometimes called "ullage."
   In shotshells such space occurs only with one-piece plastic wads. The air space is that area in and around the wad "legs."

Air Spiral—The term used to describe the corkscrew-like flight path of a bullet.

Align—To bring the front and rear sights on a gun into line with the axis of the bore.

Altitude Effect—The effect on velocity and therefore, trajectory, of a projectile or shotshell pattern caused by changes in atmospheric density due to altitude.

Ammunition—One or more loaded cartridges consisting of a primed-case propellant and with or without one or more projectiles.

Ammunition, Ball—A term generally used by the military for a cartridge with a full metal jacket or solid metal projectile.

Ammunition Code Number—A code number and/or letter(s) usually found on the carton that identifies a particular quantity of ammunition for its manufacturer.

Ammunition Color Code—A method of distinguishing the various gauges of shotshells and types of metallic ammunition by color or plating.

Ammunition, Fixed—A metallic cartridge or shotshell that is complete and ready for use.

Ammunition, Live—A cartridge or shotshell prior to being fired that is assembled with a live primer, an appropriate propellant, and projectile or shot charge.

Ammunition, Match—Ammunition made specifically for match target shooting. Produced with special controls to assure maximum uniformity of cartridge performance.

Ammunition, Metallic—A generic term for rimfire and centerfire ammunition derived from their metallic cases.

Ammunition, National Match—Ammunition produced for the American National Matches (at Camp Perry, Ohio) by appropriate government or commercial manufacturing facilities. Cartridges are usually, but not always, head stamped "NM" for identification purposes.

Ammunition, Reference—Ammunition used in test ranges to calibrate test barrels and other velocity and pressure measuring equipment.

Ammunition, Small Arms—A military term for ammunition for firearms with bores not larger than one inch.

Angle of Departure—The angle formed between a horizontal line and the center of the bore at the moment the projectile leaves the muzzle of the gun.

Angle of Elevation—The vertical angle formed between the line of sight to the target and the axis of the barrel bore.

Annulus—The ringlike space between the top of the primer and the primer pocket.

Backthrust—The force exerted on the breech block by the head of the cartridge case during propellant burning.

Ball, Patched—For modern ammunition the term patched ball refers to a full metal jacketed bullet.

Ballistics—The science of projectiles in motion. Usually divided into three parts:
- Interior Ballistics
- Exterior Ballistics
- Terminal Ballistics

Ballistic Coefficient—The ratio of the sectional density of the projectile to its coefficient of form (sometimes referred to as its Form Factor). It is an index of the manner in which a particular projectile decelerates in free flight due to the resistance or drag of the atmosphere in which it is traveling.

Ballistics, Exterior—The branch of Applied Mechanics which relates to the motion of a projectile from the muzzle of the gun to the target.

Ballistics, Interior—The science of ballistics dealing with all aspects of the combustion phenomena occurring within the gun barrel, including pressure development and motion of the projectile through the bore of the firearm.

Ballistics Table—A descriptive and performance data sheet on ammunition. Information usually includes: bullet weight and type, muzzle velocity and energy, velocity, energy and trajectory data at various ranges.

Ballistics, Terminal—That branch of ballistics which deals with the effects of projectiles at the target.

Barrel Erosion—The wearing, or physical degradation of the bore or chamber of a firearm, caused by hot powder gases or projectile passage.

Barrel Pressure—The pressure in a barrel developed by the propelling gases.

Barrel, Pressure—A heavy walled barrel fitted with instrumentation to measure pressure.

Barrel Time (Ignition Barrel Time)—The elapsed time from the contact of a firing pin with a cartridge primer to the emergence of the projectile(s) from the muzzle of the firearm.

Barrel Vibration—The oscillations of a barrel as a result of firing.

Barrel Walking—A gun barrel which, due to internal stresses, changes its center-of-impact point when heated by firing.

Barrel Whip—The movement of the muzzle end of a barrel that occurs as the projectile leaves.

Base, High—A term erroneously applied to a shotshell with a high metal cup. Properly applied it refers to the height of the internal base wad.

Battery Cup—A flanged metallic cup used in shotshell primer assemblies that provides a rigid support for the primer cup and anvil.

BB—The designation of spherical shot having a nominal diameter of .180 inch used in shotshell leads. The term BB is also used to designate steel or lead air rifle shot of .175 inch diameter. Although the two definitions cause some confusion, they have co-existed for many years.

BB Cap—The abbreviation for Bulleted Breech Cap. Originally designed as a rimfire cartridge for use in Flobert rifles in France for indoor shooting.

Big Bore—In America, any firearm using a centerfire cartridge with a bullet .30 inch or larger in diameter.

Blow Back—In ammunition, blow back refers to a leakage of gas rearward between the case and chamber wall from the mouth of the case.

Body (Case)—The portion of the metallic cartridge case that contains the propellant. In shotshells it is the tubular section that contains the propellant, wad, and shot charge.

Bore Sighting—A method of aligning a barrel on a target by aiming through the bore. May be part of the sight-alignment procedure.

Brass, Low/High—Common terminology referring to the length of the external metal cup on a shotshell. Properly called "low cup" or "high cup."

Brisance—A term describing the shattering power of high explosives.

Buckshot—Lead pellets ranging in size from .20 inch to .36 inch diameter normally loaded in shotshells.

Bullet—A non-spherical projectile for use in a rifled barrel.

Bullet, Boattail—A specific design of bullet having a tapered or truncated conical base.

Bullet, Capped—Consists of a standard lead-type bullet having a harder metal jacket (gilding metal, copper, etc.) over the nose.

Bullet, Cast—A bullet formed by pouring molten lead alloy into a mold.

Bullet, Copper Jacketed—A bullet having an outer jacket of copper or copper alloy, and containing a lead alloy core.

Bullet Core—The inner section of a jacketed bullet.

Bullet Creep or Popping—The movement of a bullet out of the cartridge case due to the recoil of the gun and the inertia of the bullet. Also called bullet starting.

Bullet Diameter—The maximum dimension across the largest cylindrical section of a bullet.

Bullet, Flat-Nosed—A bullet with a flattened front end at right angles to its axis.

Bullet, Expanding—A hunting bullet design that provides for controlled expansion on impact.

Bullet, Frangible—A projectile designed to readily break up upon impact on a hard surface in order to minimize ricochet or spatter.

Bullet, Full Metal Jacket—A projectile in which the bullet jacket encloses most of the core with the exception of the base.

Bullet, Gas Check—A lead-alloy bullet with a copper or gilding metal cup pressed over the heel.

Bullet, Hollow Base—A bullet with a deep heel cavity.

Bullet, Hollow Point—A bullet with a cavity in the nose to facilitate expansion.

Bullet Jacket—A metallic cover over the core of the bullet.

Bullet Mushroom—A bullet that has expanded upon impact to a mushroom-like shape.

Bullet Ogive—The curved forward part of a bullet.

Bullet, Round-Nose—An elongated projectile with a radiused nose.

Bullet, Semi-Wadcutter—A projectile with a distinct shoulder and short truncated cone at the forward end.

Bullet, Soft-Point—A design providing for exposure of a portion of the cone at the nose of a jacketed bullet.

Bullet Spin or Rotation—The rotational motion imparted to a bullet by the rifling in the barrel.

Bullet, Spitzer—A bullet design having a sharp-pointed, long ogive, usually of seven calibers or more (i.e., length to diameter ratio).

Bullet Splash—The spatter and fragmentation of a bullet upon impacting a hard surface.

Bullet Stabilization—The act of steadying a bullet in flight by the use of the proper rifling, twist, and bullet velocity.

Bullet, Steel Jacketed—Plated or clad steel is sometimes used as a substitute for gilding metal or copper for bullet jacket material.

Bullet, Swaged—A bullet that has been formed by impacting the bullet material into a die.

Bullet Tipping—The instability of a bullet in flight.

Bullet, Truncated—A design of a flat-nosed bullet having a conical shape rather than a nose formed by a radius.

Bullet Upset—In interior ballistics: The expansion or slugging of the heel of a bullet under the initial acceleration when the bullet first contacts the rifling. In exterior ballistics: The expansion of the nose of a bullet upon impact of target.

Bullet, Wadcutter—A generally cylindrical bullet design having a sharp shouldered nose intended to blank target paper to facilitate easy and accurate scoring.

Bullet, Wax—A bullet made from paraffin and other wax preparations, usually used for short-range, indoor target shooting.

Bullet Wobble—A characteristic caused by the eccentricity or imbalance of the bullet to the axis of the bore.

Caliber—A numerical term in a cartridge name to indicate a rough approximation of the bullet diameter.

Cannelure—A circumferential groove generally of corrugated appearance cut or impressed into a bullet or cartridge case.

Cap—An obsolete term referring to a primer.

Cap Flash—The ignition of a primer produces a high temperature flash of hot gases, which is called the "Cap Flash" or "Primer Flash."

Cartridge—A single round of ammunition consisting of the case primer and propellant with or without one or more projectiles.

Cartridge, Big Bore—For target matches in the United States, cartridges using bullets .30 inch or more in diameter.

Cartridge, Bottleneck—A cartridge case having a main body diameter and a distinct angular shoulder stepping down to a smaller diameter at the neck portion of the case.

Cartridge Cook-Off—The firing of a cartridge by overheating in a firearm chamber. Usually associated with machine guns.

Cartridge, Definitive Proof—A cartridge loaded to specified pressures higher than service loads. Used only for testing assembled firearms.

Cartridge, Magnum—A term commonly used to describe a rimfire or centerfire cartridge, or shotshell, that is larger, contains more shot, or produces higher velocity than standard cartridges or shells of a given caliber or gauge.

Cartridge, Neck of—The reduced diameter cylindrical portion of a cartridge case extending from the bottom of the shoulder to the case mouth.

Cartridge, Rebated—A centerfire cartridge case of which the head is smaller than the diameter of the body of the case.

Cartridge, Rimless—A centerfire cartridge whose case head is of the same diameter as the body and having a groove turned forward of the head to provide the extraction surface.

Cartridge, Small Bore—General term applied in the United States to rimfire cartridges. Normally used for target shooting.

Cartridge, Wildcat—Cartridges that have never been commercially manufactured and made publicly available.

Case—Refers to cartridge case. Shortened through common usage.

Case, Belted—A cartridge-case design having an enlarged band ahead of the extractor groove. This type of construction is generally used on large capacity, magnum-type cartridges.

Case Capacity—The amount of weight of a *particular type of powder* that can be inserted into a cartridge case with the bullet fully seated, without compressing the powder charge.

Case Life—An expression of the number of times a case can be reloaded and fired.

Case Mouth—The opening in the case into which the projectile or shot is inserted.

Case Mouth Chambering—A reaming operation performed on cartridge cases prior to reloading to provide a taper at the case mouth for ease of bullet seating.

Case Shoulder—The section of a bottleneck cartridge case connecting the main body of the case and the smaller diameter neck.

CB Cap—A low-velocity .22-caliber rimfire cartridge having a conical bullet loaded in a case shorter than the 22 Short.

Center of Impact—The center of a shot pattern or target made by a series of rounds fired at the same aiming point.

Charge—The amount, by weight, of a component of a cartridge (i.e., priming weight, propellant weight, shot weight).

Charge, Maximum—The heaviest charge weight, in grains, of a particular propellant that may be used with other specified ammunition components without exceeding the safe, maximum, allowable pressure limit for the specific cartridge or shell being loaded.

Chronograph—An instrument designed to measure elapsed time. When correlated with the proper tables it can be used to indicate projectile velocity. Most of the newer units are designed for direct velocity read-out.

Coefficient of Form—A numerical term indicating the general profile of a projectile.

Crimp, Rolled—The closure of the mouth of a shotshell by inverting the mouth of the tube over a top wad or slug.

Crimp, Star—A type of closure of the mouth of a metallic case or shotshell in which the sidewalls are folded in a star-shaped pattern.

Deflection—The variation in the normal flight path of a projectile caused by wind or other external influences.

Delayed Fire—Any delay in firing of an abnormal duration. This implies that firing does eventually occur.

Disc, Paper—A small circular piece of treated paper cut and pressed into the primer cup in contact with the primary mixture.

Dispersion, Horizontal or Vertical—The greatest horizontal or vertical distance between any two bullet holes on a target, normally measured center to center.

Doubling—The unintentional firing of a second shot.

Dram Equivalent—The accepted method of correlating relative velocities of shotshells loaded with smokeless powder to shotshells loaded with black powder. The reference black-powder load chosen was a 3-dram charge of black powder, with 1⅛ ounces of shot with a velocity of 1200 fps. Therefore, a 3-dram *equivalent* load is one using a charge of smokeless powder that pushes 1⅛ ounces of shot to a muzzle velocity of 1200 fps. For a complete listing of dram equivalents refer to Chapter Three, "Recoil Calculation, an Important Step in Cartridge or Shell Selections." Modern ammunition smokeless-powder charges are drastically lighter than the dram-weight charges of the old black-powder cartridges. Never load shells using a dram-weight charge of smokeless powder.

Energy Formula—The following formula is used to obtain the kinetic energy of a projectile in foot-pounds:

$$K.E. = \frac{WV^2}{450{,}240}$$

W = Weight of the bullet in grains
V = Velocity of the bullet in fps

Five-In-One Blank—A blank cartridge that was designed for use in firearms of three different calibers. These calibers are 38-40, 44-40, and 45 Colt. The Five-In-One designation was most likely a promotional man's pitch and is based on .38-40 and .44-40 applications in rifles and handguns.

Flash Hole—1. A hole pierced or drilled through the center of the web in the primer pocket in a metallic cartridge case. 2. The hole in the end of a battery cup primer used in shotshells.

Flash Inhibitor—A material that is added to propellant for the purpose of reducing muzzle flash.

Fusing—1. The balling of lead shot due to gas leakage. 2. The melting of the core of a jacketed bullet. 3. The melting of a lead-alloy bullet.

Gauge—A term used in the identification of a shotgun bore. Related to the number of bore-diameter lead balls that it takes to weigh one pound.

Grain—A unit of weight (avoirdupois), 7,000 grains per pound.

Group Measurement—The determination of the center to center distance between the two bullet holes farthest apart on a target. This is referred to as the group extreme spread.

Hardball—A slang term for full-metal jacketed 45 ACP pistol ammunition.

Head—The end of the cartridge case at which the primer is inserted.

Heel—The rear portion of a bullet.

Heel Cavity—A recess in the base of a bullet.

High Brass—Common terminology referring to the length of the external metal head on a shotshell. Properly called "High Cup."

Hull—A slang term for a cartridge or shotshell case.

Jacket—The envelope enclosing the lead core of a compound bullet.

Keyhole—An oblong or oval hole in a target that is produced by an unstable bullet striking the target at an angle to the bullet's longitudinal axis.

Leading—The accumulation of lead in the bore of a firearm from the passage of lead shot or bullets.

Load, Brush—A shotshell load specifically designed to provide a widely spread pattern at a close range in a choked gun. Also called a "Spreader Load," "Scatterload," "Bush Load," and "Thicket Load."

Load, Duplex—A cartridge case containing two projectiles.

Load, Field—A shotshell loaded for hunting small-game animals and birds.

Load, Squib—A cartridge or shell which produces projectile velocity and sound substantially lower than normal. May result in projectile and/or wads remaining in the bore.

Loading Density—The relationship of the volume of the propellant to the available case volume. Usually expressed as a percentage.

Metal Fouling—Metallic bullet material deposited in the bore of a firearm after the firing of one or more cartridges.

Midrange—1. A term that defines a specific point in the trajectory of a projectile that is half the distance between the firearm and a target. 2. A reduced velocity, centerfire cartridge, used principally in target shooting.

Minute of Angle (M.O.A.)—An angular measurement method used to describe accuracy capability. A minute of angle is one sixtieth of a degree and subtends 1.047 inches at 100 yards whch, for shooting purposes, is considered to be one inch. A minute of angle group, therefore, subtends one at 100 yards, two inches at 200 yards, and so on.

Misfire—A failure of the priming mixture to be initiated after the primer has been struck an adequate blow by a firing pin or the failure of the initiated primer to ignite the powder.

Mushroom—A descriptive term for a bullet that is designed to expand to a larger diameter upon impact with a target.

Muzzle Blast—The resultant noise that occurs at the muzzle of a firearm when the projectile leaves the muzzle and the hot gases are released.

Muzzle Wave—The air that is compressed and moves out radially from a firearm's muzzle after firing a projectile.

Nitroglycerin—The common term for nitric acid ester of glycerin, i.e., glyceryl nitrate. It is incorporated with nitrocellulose smokeless-powder formulations to make double base powder.

Noncorrosive—A term applied to primers that contain no chemical compound that could produce corrosion or rust in gun barrels.

Nose—The point or tip of a bullet.

Ogive—The curved portion of a bullet forward of the bearing surface.

Ogive, Secant—A projectile nose with the curvature not tangent to the cylindrical bearing.

Oilproof—The treatment of a cartridge to minimize the entry of oil or water.

Open Point—A type of bullet having a cavity in the tip.

Over-Bore Capacity—A cartridge which contains more powder than can normally be burned in that bore diameter and volume.

Pattern—The distribution of shot fired from a shotgun. Generally measured as a percentage of pellets striking in a 30-inch circle at 40 yards.

Pellet (shot)—A common name for the small spherical projectiles loaded in shotshells.

Plinking—The informal shooting at inanimate objects located at arbitrary or indefinite distances from the firing point.

Point of Aim—The exact point on which the shooter aligns the firearm's sights.

Point of Impact—The point at which a bullet hits a target.

Powder—Commonly used term for the propellant in a cartridge or shotshell.

Powder-Burning Rate—The speed with which a propellant burns. It is affected by both physical and chemical characteristics.

Powder Deterioration—The chemical decomposition of modern smokeless propellants, usually occurring over a long period of time, combined with improper storage conditions.

Powder, Double-Base—A propellant composed of colloided nitrocellulose and nitroglycerin as its base as opposed to a single-base powder which has only colloided nitrocellulose as its base material.

Powder Fouling—Powder residue left in firearms after firing.

Pressure—The force developed in the cartridge (and gun) by the expanding gases generated by the combustion of the propellant.

Pressure Curve—A graph of the relationship of chamber pressure to time or travel in a firearm when a cartridge or shell is fired.

Pressure, Residual—The pressure level that remains in the cartridge case or the shell within the firearm's chamber and in the bore, after the projectile leaves the muzzle of the firearm.

Primer—An ignition component consisting of brass or gilding metal cup, primaring mixture, anvil, and foiling disc.

Projectile Rotation—The spinning motion that is imparted to a projectile due to engagement with the rifling in the barrel of a firearm, as it is driven down the barrel. The rate of spin rotation is dependent upon the rate of twist of the rifling and the velocity. The barrel twist (left or right) determines the direction of the rotation.

Projectile, Sabot-Type—A sub-caliber projectile centered in a light-weight carrier to permit firing the sub-caliber projectile in a larger-caliber firearm.

Range, Effective—The maximum distance at which a projectile can be expected to be lethal.

Range, Maximum—The greatest distance a projectile can travel when fired at the optimum angle of elevation of the gun barrel.

Rate of Twist—The distance required for the rifling to complete one revolution.

Recoil Pendulum—A device for measuring Free Recoil Energy in which a firearm is suspended from fixed points so as to allow it to swing freely while the barrel remains horizontal.

Ricochet—The glancing rebound of a projectile after impact.

Sectional Density—The ratio of a bullet weight to its diameter.

Shocking Power—A colloquial term used to describe the ability of a projectile to dissipate its kinetic energy effectively in a target.

Shock Wave—The disturbance of air surrounding and behind the bullet caused by a compression of the air column directly in front of the bullet.

Shot, Balled—The fusing together of several pellets in a shotshell load, usually caused by hot propellant gases leaking past the wadding and fusing the shot while the shot is still in the barrel.

Shot, Bird—A general term used to indicate any shot smaller than buckshot.

Shot, Chilled—Lead shot containing more than 0.5% alloying metal to increase its hardness, also called hard shot.

Shot, Drop—Lead shot containing less than 0.5% alloying metal. Also called soft shot.

Shot, Dust—Lead shot having a nominal diameter of .040 inch or smaller.

Shot, Steel—Soft steel pellets made specifically for use in shotshells.

Skid Marks—Rifling marks formed on the bearing surface of bullets as they enter the rifling of the barrel before rotation of the bullet starts.

Slug, Rifled—A single projectile with spiral grooves and hollow base, intended for use in shotguns.

Small-Bore—In America, any firearm or ammunition of the rimfire type with a lead-alloy bullet not over 0.23 inch diameter.

Time of Flight—The total elapsed time that a projectile requires to travel a specific distance from the muzzle.

Tipping—The flight characteristic of a projectile resulting in its entering the target at an angle to the trajectory.

Trajectory—The curved path of a projectile from muzzle to target.

Trajectory, Midrange—The distance, measured in inches, that a projectile travels above the line of sight at a specific point in the trajectory that is half the distance between the firearm and the target.

Twist—The distance required for one complete turn of rifling usually expressed as a ratio—i.e., 1 in 10″.

Web—1. The solid portion of a brass centerfire cartridge case between the inside of the case at the head end and the bottom of the primer pocket. 2. The smallest dimension of a smokeless-powder granule.

Yaw—The angle between the longitudinal axis of a projectile and a line tangent to the trajectory at the center of the projectile.

CHAPTER EIGHTEEN

Shooting Safety

This book has been devoted to ammunition, and the author has tried to refrain from discussions on firearms or any other associated topic. However, he would be remiss if the topic of shooting safety was not covered.

There are a number of musts for shooting safety. Safety glasses are perhaps the most vital. Shooting glasses or prescription glasses with full safety lenses should be worn whenever shooting, regardless of the gun or ammunition used. Modern ammunition is perhaps one of the safest and most reliable consumer products available, but no one is infallible. Mistakes sometimes occur and can cause some startling results. Shooting glasses can keep these surprises in the realm of something less than a major accident. Eyes are extremely difficult to replace.

Ear protection, of course, is the second most important item for self-protection. Almost every old-timer in the shooting industry suffers from hearing loss, and the range varies from a modest loss to almost being stone-deaf. Hearing protection was almost unheard of 25 or more years ago. As a result, there are thousands of people today who can attest to the stupidity of not wearing hearing protectors.

A good pair of "ear muffs" are an essential part of any shooter's kit. Granted few will wear such devices while hunting, but the few shots fired in an average day's hunt won't rob your hearing to any noticeable degree. However, hearing protection should be worn at all times when on the range or when shooting skeet and trap. Sighting-in sessions also demand hearing protection.

The National Shooting Sports Foundation lists the rules of shooting safety as follows:

1. Treat every gun with the respect due a loaded gun.
2. Be sure of your target before you pull the trigger.

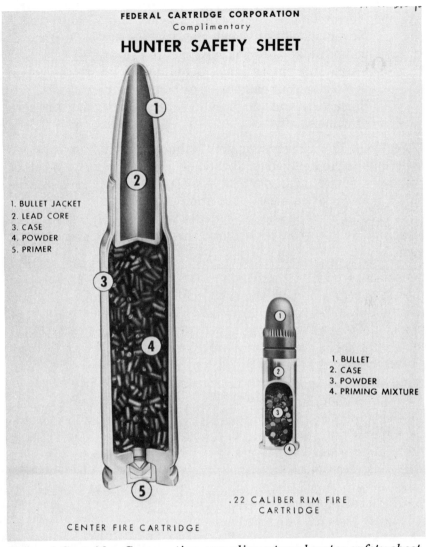

HUNTER SAFETY SHEET

1. BULLET JACKET
2. LEAD CORE
3. CASE
4. POWDER
5. PRIMER

1. BULLET
2. CASE
3. POWDER
4. PRIMING MIXTURE

.22 CALIBER RIM FIRE
CARTRIDGE

CENTER FIRE CARTRIDGE

Federal Cartridge Corporation complimentary hunter safety sheet.

3. Always be sure that the barrel and action are clear of obstruction.
4. Never point your gun at anything you do not want to shoot.
5. Never leave your gun unattended unless you unload it first.
6. Avoid alcoholic beverages both before and during shooting.

7. Never climb a tree or cross a fence with a loaded gun.
8. Never shoot at a hard, flat surface or the surface of water. Make sure you have a safe backstop.
9. Carry only empty guns, taken down or with the action open, into your camp, car, or home.
10. Store guns and ammunition separately beyond the reach of children.

Accidents don't always happen to the other guy. Firearms can be lethal instruments when abused or used improperly. Whether you are a novice or an old pro, a heartbreaking disaster can be only a moment of careless inattention away.

Be careful and the shooting sports will give you a lifetime of pleasure. Be less than careful and the lifetime lost may be your own.

Index

Accelerator cartridge 163
Accuracy with rimfire cartridges 6
Ammunition
   listing of *see* Cartridges, Shotshells, New
      products 153
Angle of elevation *see* Exterior ballistics
      tables
Antipersonnel cartridges 34,79

Ballistic coefficient *see* Exterior ballistics
      tables
Ballistics tables
   see Exterior ballistics tables conditions
      pertaining to 13,49
Barrel life 27,29,30,31,32,38,139
Base wad 147
Big game 26
Black powder loads 38
Blank cartridges 176
Bullets 146
Bullet drop *see* Exterior ballistics tables
Bullet jacket fouling 27
Bullet seating 150

Calculating recoil 17-18
Cannelure 150
Cartridges
   Centerfire Handgun
     .22 Remington Jet 81
     .221 Remington Fireball 81
     .25 Auto 79
     .256 Winchester Magnum 81
     .30 Luger 80
     .32 Short Colt 80
     .32 Long Colt 80
     .32 Colt New Police 80
     .32 S&W 80
     .32 Auto 80
     .32-20 80
     9mm Luger 82,142
     9mm Winchester Magnum 172
     .357 Magnum 81,142
     .38 Short Colt 80
     .38 Long Colt 81
     .38 Colt New Police 81
     .38 S&W 81
     .38 Special 82,142,156

.38 Auto 84
.38 Super Auton 84
.380 Auto 84,161
.38-40 85
.41 Remington Magnum 85
.44 S&W Special 85
.44 Remington Magnum 85,156,165
.44-40 Winchester 86
.45 Auto 86
.45 Auto Rim 86
.45 Winchester Magnum 172
.45 Colt 87

Centerfire Rifle
   .17 Remington 27
   .218 Bee 27
   .22 Hornet 27,138
   .22 Savage 28
   .222 Remington 28, 139
   .222 Remington Magnum 28
   .223 Remington 28
   .225 Winchester 28
   .22-250 Remington 29,139,156
   .243 Winchester 29,139,156
   6mm Remington (.244 Remington)
     29,155
   .25-06 Remington 30
   .25-20 Winchester 30
   .25-35 Winchester 30
   .250 Savage (.250-3000) 30,139
   .256 Winchester 31
   .257 Roberts 31
   6.5mm Mannlicher-Schoenauer 31
   6.5x55mm (6.5 Swedish) 32
   6.5mm Remington Magnum 32
   .264 Winchester Magnum 32
   .270 Winchester 32,156
   7mm Mauser (7x57mm) 3,140
   7mm Remington Magnum 33,156
   .280 Remington 33
   .284 Winchester 34
   .30 Carbine (.30 M-1 Carbine) 34
   .30 Remington 34
   .30-30 Winchester 34,140
   .300 H&H Magnum 35
   .300 Winchester Magnum 35
   .30-06 Springfield 35,163

.30-40 Krag 36
.300 Savage 36
.303 Savage 37
.308 Winchester 37,140
.303 British 37
8mm Mauser (8x57mm) 38
8mm Remington Magnum 164
.32 Remington 38
.32 Winchester Special 38
.32-20 Winchester 38
.32-40 Winchester 38
.338 Winchester Magnum 39
.348 Winchester 39
.35 Remington 39
.350 Remington Magnum 39
.357 Winchester Self-Loader 40
.358 Winchester 40
.375 Winchester 167
.375 H&H Magnum 41,141
.38-55 Winchester 41
.38-40 Winchester 41
.44-40 Winchester 41
.44 Remington Magnum 41
.444 Marlin 41
.45-70 Government 42,54,142
.458 Winchester Magnum 42,142

Rimfire
    5mm Remington Magnum 2,7
    .22 CB Short 2
    .22 CB Long 2
    .22 Short 3
    .22 Long 4
    .22 Long Rifle 4,138
    .22 Xpediter 169
    .22 WMR 6

Cartridge designation 26
Cartridge/gun combination selection 24
Cartridge history 26
Cartridge interchange ability 129-130,132
Cartridge popularity 25
Cartridge selection 17,27-42,81,138
Cases 147
Centerfire rifle cartridge selection 25,27-42
Chamber erosion 3,129
Chamberlin Cartridge and Target Company
    160
Charge weight, average 150
Choke expansion 111
Chronograph 19
Components, reloading 146-148
Contamination of ammunition by oil or
    solvents 182
Cordite powder 38
Cracked cases 147
Crimping 150

Dangerous arms and ammunition
    combinations 133-135

Dangerous game 26
Decapping 149
Diamond Pyramid Hardness test 110
Disintegrating bullets 3
Disposal of ammunition 177
Double charge 150
Dram equivalent 116
    chart 117-119
du Pont de Nemours and Company 159

Ear protection 214
Effective velocity of powder gases 19
Explosives 178
Exterior Ballistics Tables
    Centerfire Handgun
        Bullet drop 94,95
        Energy 91,92,93,94,95
        Long range exterior ballistics 94
        Maximum distance to first point of
            impact 39
        Midrange trajectory 91,92,93,94,95
        Velocity 91,92,93,94,95
    Centerfire Rifle
        Ballistic coeffient 52-53,57-58,62-63,68
        Bullet drop 52-53,57-58,62-63,68
        Energy 53-54,58-59,64-65,69
        Hold-under 54-55,60-61,65-66,70
        Mid-range trajectory 54-55,60-61,65-
            66,70
        Time of flight 53-54, 58-59,64-65,69
        Trajectory 56-57,61-62,66-67,71
        Velocity 53-54,58-59,64-65,69
        Wind drift 54-55,60-61,65,66,70
    Old Cartridges see Historical exterior
        ballistics
    Rimfire Handgun 16
    Rimfire Rifle
        Angle of elevation 15
        Ballistic coefficient 15
        Bullet drop 14
        Hold-under 14
        Maximum range 15
        Midrange trajectory 14
        Muzzle velocity 15
        Time of flight 14
        Trajectory 14
        Wind drift
    Shotgun
        Distance to first impact of leading pellet
            127-128
        Drop 122,125,126
        Energy per pellet 121,123,124
        Time of flight 122,123,124
        Velocity 121,123,124
Extraction problems 3,81,129

Federel Cartridge Corporation 153-158
Fire, exposure of ammunition to 176-177
Fit, or gun 8
Flash hole 147,149

Flinching 17,32,86
Fouling, bullet jacket 27
Free recoil tables 21-23
Functioning, .22 CB Long 4

Glossary 202-213
Gunshyness 17
Gun weight 99
  vs. recoil 23

Handgun Cartridge selection 81
Handling ammunition 173
Handloading safety 151-152
Height of trajectory see Exterior ballistics
  tables
Henry rifle 1
Historical exterior ballistics tables
  Energy 191-196
  Free recoil 191-196
  Penetration 189-196
  Sighting-in 197,198-201
  Trajectory 189-196
  Velocity 189-196
History of cartridges 26
History of rimfire cartridges 1
Hold-over see Exterior ballistics tables
Hold-under see Exterior ballistics table

Ignition/barrel time 18
Immersion in water of ammunition 175
Incineration of ammunition 177
Indoor cartridges 4
Industry adjustments concerning nominal
  velocities 57
Interchangeable names 131-132
Interpolation of data 148

Lake City Army Ammunition Plant 160
Lever action rifle 25
Light big game 26
Loading operations 149
Long range ballistcs see Exterior ballistics
  tables
Long range waterfowling 96,97
Loose primer pocket 147
Low temperatures and velocity loss 97

Mass explosion 174
Match cartridges 1000 yard 35
Maximum range see Exterior ballistics tables
Mid-range trajectory see Exterior ballistics
  tables
Military ammunition
  79,80,82,84,87,153,154,160
Misfires 175
Model 1866 Winchester 1
Model 1894 Winchester 167
Muzzle blast 24,30,31,83,85
Muzzle energy see Exterior ballistics tables
Muzzle flash 84

Muzzle velocity see Exterior ballistics tables

National Shooting Sports Foundation 214
New ammunition products 153
New York Police Department 86
Noise level 7
Nomenclature 131
Nominal muzzle velocity see Exterior
  ballistics tables
Non toxic shot 112

"Oilproof" ammunition 183
Olin corporation 166
Organic solvents, exposure to of ammunition
  175
Overall case length 149
Overcharging see powder charging

Pattern density 107
Pellet deformation 160
Perceived recoil 18
Pierced primer 84
Pinholes 147
Plinking 3,6,28,34,79
"Plus P" loads 83,84,131,156
Police Firearms 84,87-88
Police Cartridges 83,85,87
Powder charging 150
Powder height gauge 150
Powder gases, effective velocity of 19
Powder, storage of 148
Practice, cartridges for 2
Primers, storage of 147
Primer dust 149
Primer pockets 149

Range limitations 26
Rebated cartridges 34
Recoil 17,19
  calculation of 19-21
  comfortable levels
  excessive 41,43
Recoil pad 8
Recoil tables 21-23
Reloading ammunition 145
Remington Arms Company, Inc. 159-166
Resizing 149
Rifling Twist 4
Rimfire ammunition see cartridges
Ruger Hawkeye 31,81
Ruptured cases 132

SAAMI see Sporting Arms and Ammunition
    Manufacturers' Institute
Sabot encased bullet 163
Safety glasses 150,214
  use of when reloading 151-152
Safety suggestions 173,214-216
Scales 148
Scuffed shotshells 186

Service cartridges 32,34,36,37,38
Shipment of sporting ammunition 173
Shot 148
  sizes 102
  charts 106-107,113-115
  steel 108,111,112
Shot charge weights 107
Shotshells
  10 gauge 3½" Magnum 96,158,171
  10 gauge 2¼" 97
  1 gauge 3" Magnum 97,158,160
  12 gauge 2¼ 98
  16 gauge 2¼ 99
  20 gauge 3" Magnum 99
  20 gauge 2¼" 100
  28 gauge 2¼ 100
  .410 bore 3" 100
  .410 bore 2¼" 101
Shotshell selection 96,142-144
Sight height *see* Exterior ballistics tables
Silhouette shooting, handgun 89
Small game cartridges 5
Smokeless powder cartridges, development of
  167
Spencer carbine 1
Spitzer bullets 39
Split cases 132, 147
Sporting Arms and Ammunition
  Manufacturers' Institute
  148,173,175,176,179,202
Squid shots 175
Stock design 8
Steel shot *see* Shot
Storage of ammunition 173-178,179-187

Storage period maximum 179
Stretch marks 147
Suburban hunting 27
Swelling of paper shotshells 186

Temperature, effect of on ammunition 180
Time of flight *see* Exterior ballistics tables
Trajectory, height of *see* Exterior ballistics
  tables
Trapping, cartridges for 2
Trimming cases 147,149
Tubular magazines 150
Twin Cities Army Ammunition Plant 153

Union Metallic Cartridge Company 159
Unservicable ammunition 177
U.S. Department of Transportation 174,177-
  178

Varmint 26
Varmint handgun 31
Varmint hunting vs. varmint shooting 27
Velocity, average *see* Exterior ballistics tables
Velocity, shotshells 108
Velocity variations 45
Vented test barrel 89
Volcanic pistol 1

Wads 148
Western Cartridge Company 167
Wildcat Cartridges 28,29,30,152
Winchester Repeating Arms Company 167
Winchester-Western 160,166
Wind drift *see* Exterior ballistics tables